ULTRA HI DEF MARKETING

5-Step Guide to World
Domination in the Tech
Industry

Coleen Sterns Leith

Ultra High Def Marketing by Coleen Sterns Leith

ISBN # 978-0-9982239-8-8

Edited by Eli Gonzalez

Book Design by Marketing Matters

Printed in USA

Dedication

To my husband, Scott who encouraged me and never doubted.

Introduction

"If passion drives you, let reason hold the reins."

– Benjamin Franklin

I love what I do.

I love creating marketing campaigns. I love that I never know what my day will bring. I love how rapidly marketing tools develop and evolve. I love data and technology. As a marketer who specializes in the technology and AV industry, I really could not be better suited in another career. Each day, I get to collaborate with other creative and passionate people to implement, test, measure, or fine-tune a company's message to the world. It's a huge undertaking, being responsible for how a company is portrayed in the marketplace, and one I take very seriously.

I love to help people. I'm a nurturer at heart, a tough-nosed and opinionated one. In college, my intent was to become a nurse. I had participated in a nursing co-op program during high school and figured I could help people that way. I graduated high school and entered college to obtain a nursing degree. When I wasn't learning about biology and other sciences, I worked at a local hospital to get some hands-on experience. After several years, I realized that no matter how hard I wanted to help my patients, there was only so much I could do. I didn't like not having more control of their outcome. I knew I would be unhappy being a nurse so I discontinued my pursuit of a nursing degree, switched my major to business and marketing then eventually went to work for my father in southern California.

His company, Xenotronix, engineered and manufactured custom battery chargers and power supplies for the US government and the entertainment industry. My father placed me in technical sales and I quickly found out why; I had a natural propensity for it. As my aptitude increased, I moved on and along the way I found my home in the consumer electronics industry. It was fast, competitive, loud, and constantly morphing. After about seven years in the CE industry, I was having lunch with our print salesman at the time, and dear friend Dennis Weddington, he asked me how long was I going to make other people look good from my ideas and work ethic. "Why don't you do what you do for yourself?", he suggested.

By the time the check was paid, he had offered me very favorable credit terms with our key printing partner if I started my own business. Essentially, he allowed me to start my own business without prepaying for jobs or going through the frustrating rigors of establishing credit for the business. Marketing Matters was born.

That was over twenty years ago and I'll admit, if I knew how much I didn't know at the time I probably would not have struck out on my own. Yet, I had been working in the consumer electronics industry for seven years and had established myself with some major players and had cultivated good working relationships. I loved being a business owner. I had found a job where my direct efforts had a huge part in my company's outcome.

The first couple of years weren't rosy but I doubled down and invested as much as I could back to the business. My big break happened in 1998 when a start-up, AVAD, hired Marketing Matters. I hired my first employees and we were off and running. I learned how to manage the business and eventually joined Vistage, a CEO group, and learned a lot from that experience; in particular, the importance of belonging to a community of like-minded individuals.

Today, I love to work with established brands and start-ups alike. I was once a start-up and humbly I'll say that Marketing Matters has established itself with some of the biggest players in our space. I still love being a part of the consumer electronics and pro AV industry.

Over the years, we have fine-tuned our marketing methodologies. We have seen many marketing companies enter our space only to enter another one in a few years. To be successful in consumer electronics, you can't fake it. You must understand the players, the products, the consumers and B2B buyers.

We have many testimonials and success stories. We helped one start-up technology company go from a start-up to a company grossing over $200 million in sales in just seven years. We have won over 60 design and publishing awards thus far, far more than just about every other company in our space. We have been sought after by major brands such as Bose and Sharp. Our vision is not for our clients to be profitable; we want more for them – Total World Domination.

The U.S. Census Bureau reports that 400,000 new businesses are started every year in the United States. The troubling part is that they estimate over 470,000 new businesses are dying each year. At this rate, more companies are closing than are being opened. The reason why so many companies are in trouble is simple – they don't have enough sales or are unable to manage their expenses. Sales, with good expense management, is the "cure-all" for any business. You can survive bad management, a poor process, or even a fire if you are making enough money. However, even with great management, a great process, and a clean warehouse, you will go out of business without enough sales.

One of the main reasons why companies struggle with sales is because they don't know how to let enough of the market know what they do or the benefits of their product/service. That's where we come in. Our formula works.

Yet, for as many companies as we help, more close their doors. I wrote this book to give back to the industry that has been so good to me. There are great products and new visions being realized seemingly on a daily basis. As a total geek and lover of technology, I want to find them, buy them, and use them.

This book contains much of what we currently do today for our clients. You won't find fluff within these pages. I have a system – TWD (Total World Domination) and I'm presenting most of it to you here. I hope this book falls into the hands of marketing managers and business owners; people that know they have a wonderful product yet for the life of them can't get a foothold on the market.

You will find plenty of solid tips and tools for success in the consumer electronics and integrated technology industries, whether you are an industry veteran or just stepping into the industry. Passion, drive, creativity, and relentless determination will contribute to your success, but all of that will be for naught if you don't have a proven marketing strategy. You'll find it in this book.

Coleen Sterns Leith

President/Founder, Marketing Matters

Table Of Contents

Chapter 1

∽◌☙⚜❧◌∾

Marketing Matters

"Marketing is a contest for people's attention"—Seth Godin, Author

Marketing matters.

You may have the best product in the world or provide the best service in the world. But if enough people don't know about you, your business will not survive.

You may have heard this saying, "It's all about who you know." Although that may be applicable in different areas in life, it's not true when it comes to marketing your business. I say, "It's not all about who you know; it's all about who knows you."

Again, marketing matters.

Marketing is a business discipline and, as a business owner or manager, you need to understand that it's far from just a term to describe how you let others know about your products or services. Marketing is a business discipline. It literally shapes how others view and interact with your company, and it helps develop relationships with others before conversations with prospects are ever held.

Business owners should have a clear and effective marketing strategy if they want to become business leaders. When an effective marketing strategy is successfully executed, the demand for your product or service will increase, which of course will result in greater product revenue.

After working with business owners for more than twenty years, I've learned a few facts about them. One is that many folks are not clear about what marketing is. So, before we continue, let me offer a definition. BusinessDictionary.com refers to the word "marketing" as "The management process through which goods and services move from concept to the customer."

Now that we all understand what marketing is, let's dive in.

Innovative Marketing

If you don't stand out, you don't get noticed. In order to stand out, you need to be innovative in your marketing, different in your line of thinking. As much as marketing is a visual attraction, it is first a thought in someone's mind. Think for a moment about your market and your competition. How can you be different than them in order to stand out in the marketplace? This is called a *marketing position or differentiation.*

Think differently about your approach. Become the change that you want to see in the marketplace. For example, take a view of the marketplace. If people in your line of work tend to utilize red and blue colors 98 percent of the time; then create a bright purple logo.

Make them curious.

Now, of course, don't be different just to be different. You need a strategy. Think about the book covers in a Barnes & Noble bookstore. What are book covers secretly trying to accomplish as you stroll past them? Those covers have been strategically created and placed. Their sole purpose: to get your undivided attention. They are, in essence, marketing themselves.

Depending on which genre you favor, you may never make it to that aisle in the store where that genre resides. The reason is another cover might have held you at book-point and forced you to read the book blurb. Now you're hooked!

Think of the bookstore like your marketplace. How can you make an advertisement that will attract the attention of consumers, changing their minds from what they were thinking of buying to buying your product?

Again, be different. Be unique.

Marketable Attention

I have always loved to travel. When I was younger, I would

frequently take cross-country trips across the United States. There were times when I felt I was in the middle of nowhere, and I guess I was at times, such as when I'd be in the middle of the western American landscapes.

I remember one vivid image from the state of Nevada that still permeates in my mind to this day. As we drove across the black road and I gazed at the brown sandy desert terrain, everything was bland. Boring and bland. Then as I stared in the distance and I saw a mountain, but not just any mountain. It looked bright orange.

I thought, Wow, that mountain has "*life.*"

To a degree, the mountain gave me life. I'll explain. I showed pictures of that mountain to my family and friends, and they were impressed. Those images birthed in me the ability and awareness of how to influence other people by the way I presented something. People that saw the image and how I explained the feeling I had when I saw the mountain. Then they'd say they'd love to see it in person. I didn't know what marketing meant when I was younger, but I know what that orange mountain means today—marketability.

The mountain looked majestic in the distance. And while there were other mountains, they couldn't attract my attention. It was the first time I saw the importance of being able to stand out. The orange mountain taught me my first lessons about marketing: everyone notices you if you find out how to stand out. As I grew older, I realized how much I enjoyed making businesses stand out in the marketplace.

Be Visible in your Marketability

When the company AVAD first came to fruition it was the one of the first national distributors in the custom installation industry. I was fortunate to be a part of it. In 1998, one of our media strategies was to keep AVAD constantly in the news and make them a visible part of the CEDIA community.

We wanted to establish credibility in the industry for AVAD so they could attract the market-leading national lines to better serve their growing dealer base. That visibility and the favorable economy allowed us to help AVAD reach its goals. The publicity and AV community presence helped move the needle.

Focusing on our customer's win, we submitted the work we did for industry awards and won some rather surprising ones, including the 2007 Custom Publishing Council's Best New Publication Award for AVAD's "Gear Head" magazine.

It really was so amazing to be in Manhattan with all of the nation's largest publishers—think Conde Nast, Hearst Publishing, Martha Stewart, etc. To top it off, we won the top award of the year! My heart still gets a little pitter-patter when I think of that day, and the reason why is that we were able to make our customer's hearts pitter-patter. We were able to deliver amazing results for our clients and it built the foundation for a successful business for our customers and for Marketing Matters.

Influential People Can Influence your Market

As I spoke about earlier, marketing matters, in every aspect of the word. With that said, how will you matter to the market? Here's a great tip. If you can, get a very influential person to speak on behalf of the company. Amazing things could happen if you partner with the right celebrity.

I would assume that the amount of people who know about your products or service would multiply. Common sense would suggest that if you were to double the amount of people who knew about your business you can reasonably expect to double your business. Even if you didn't grow the number people who know about you, out of those who already know of you were to trust you more, based on the referral from the celebrity or the influential endorser, you should expect an increase in sales. Now, if both things were to happen; your company's audience would increase and those who know of your company would trust it more, your business could potentially skyrocket.

Not only that, but if the person is some sort of a celebrity, you gain additional exposure through their social media outlets. With consistent messaging about your company, you could easily get many of that person's followers to follow your company. In the online world of business, it is good marketing strategy to have an influential person speak on your company's behalf.

For those who don't exactly know what an influencer is, John Lincoln, CEO of Ignite Visibility, describes it perfectly:

"An influencer is someone who people listen to online. When

it comes to search engine optimization, having an influencer in your corner will mean more people link to your website, share your blog posts and trust your content. If possible, have an influencer who is a subject-matter expert head up the content creation on your website. This person can be you, someone from your company, or someone you align with."

The 100 Percent Marketing Rule

As I learned from one of my industry mentors, Joe Piccirilli, a founder of Florida's Sound Advice retail chain, AVAD and RoseWater Energy Group, there is a general rule in marketing – nothing works 100 percent of the time. Sure, a method may work for a while—until it doesn't. Think of the most iconic television ads you've ever seen. If they are older than 2 years old, it's most likely not on television anymore. Why? Because, as great as it once was, it stopped being effective. This is a copycat world. In sports, if a new coach implements a new offense that works, pretty soon all of the coaches are implementing the same style of play — at least until another coach develops a defense for it. Sound creativity is key. It's imperative to work with people who can well up a fountain of ideas to keep the company's messaging fluid.

There are many sound reasons to work with an agency as opposed to hiring in-house talent. And although I'll get more into the details on those reasons later, I'll share a few now. A marketing employee's skill set tends to be broad rather than that of a specialist. In-house staff can't provide economies of scale benefits and savings. Typically, turn around times for in-house projects are longer than what an agency provides. If that marketing employee also has other duties

– and this is common in start-ups and small companies, marketing projects may reduce the time left for those other business activities. Lastly, in-house marketing staff frequently has less experience than the agency who specializes in your industry.

I've owned a marketing agency for more than 20 years. You need to be consistently right to be in business for that long, and I'm pleased that our strategies have paid off handsomely for our clients. The reason for this book is that there are many more people we would like to assist but, either due to location or karma, we will never have the opportunity to speak to them. Yet they can read this book.

I have developed a five-step process that covers just about everything anyone needs to know to market his or her company like an expert. Although it's only a five-step process, each step is made up of multiple action items. I summed it up to five steps to make it more memorable, thus more impactful. Here are the five steps: (and pardon me if some of these words were invented . . . hey, I'm one of those creative types!)

1. Identify—Identify who you are, your audience, your goals, building a corporate identity and much more.

2. Simplify—In the tech world, the geeks have taken over the marketing copy and . . . well, that's not good. Simplifying your messaging and other parts of your business will go a long way in getting more sales.

3. Toolify—What tools are you using? Which tools should you discard and which tools should you adopt?

4. Amplify—There's a good way to get those tools and your message out there and there's a best way. The best way works better.

5. Tweakify—How do you evaluate your trade show goals, for instance? Many companies don't know how to test, measure, and adjust to what the analytics of their marketing efforts are telling them. I do.

These five steps—Identify, Simplify, Toolify, Amplify, and Tweakify are rock-solid foundations to build on to market your tech company to other companies and to the consumer masses. Some of the content in the next chapters has been tried and proven. Other parts of this book are my best forward thinking in the ever-changing landscape of marketing a technical company and its products.

I'm excited for you to read my five-step process. Your company can be a solid mountain without reading this book. However, should you read this book and incorporate much of what I recommend, you could be a bright orange mountain.

PART 1:

IDENTIFY

Chapter 2

Identify

We've all heard the adage, "Ready, Set, Go!" It's not, "Hurry Up and Go!" Many startup business owners or people that launch a new product have got it all wrong. They just hear the "Go!" They don't take the time to get Ready and then to get Set. There is a difference between getting ready and being set to then go. Right now I want to talk to you about getting ready.

As no two people are the same, no two companies are. Siblings might share a lot of characteristics, but they're not the same, not even identical twins. Sure, they may look the same from the outside but the truth is that their parents and other people close to them know which is which. And that's from the exterior. From inside, even though they may share a lot of similar likes and habits, they are very different.

If even twins are not the same, companies are even further

apart. Their processes, employees, time schedules, managers, and products are different. Like twins, even if two companies produce the same types of services or products, they are still different. Many consumers will never know how different your company is and how unique parts of it might be so they will assume that your company is no different than others in your space. Therefore it's important for you to have a company identity.

What's Your Story?

The first step in understanding your identity is to identify what makes your company tick. For starters, what's your story? How did your company get started? Every company has a point of origin. Companies are not like mountains that have always seemed to just be there. Your company has a unique story about how it began.

Did you always want to do what you are doing? Did you go to school for it? Or did life take a few turns and you found yourself starting a business? Can you remember why you wanted to start your business? These questions are essential in forming your identity. If you haven't done this yet, I recommend you sit down and answer these questions. I'm sure you have a fascinating story, and guess what, your customers want to hear it!

Every market is inundated with the buzz words 'brand loyalty'. Here's a tip — people aren't going to be loyal to your company because of the color scheme of your website. Your story, team, benefits, customer service, quality and price drive loyalty. Corporate storytelling is a vastly under-utilized tool in tech marketing that can really help drive that brand loyalty you seek.

By creating an identity for your company through storytelling, you are able to create a long term personal connection with your brand. The more compelling and unique your story, the better the connection. Your brand should be inspiring, authentic, trustworthy and creative. Let folks know what you believe in and share your vision with your followers. Listen to your customers and fine tune your story with the elements that resonate best with them. Don't sell. Tell.

So again I ask you, what's your story?

After viewing website analytics for years, we've found that the second most important page of a company website is the About Us page (right after the Contact Us page). People want to know your story. They want to know who they are doing business with. Jeremy Spark has an article on www.rocketspark.com in which he says, "So don't just copy and paste your mission statement on your page (About Us). That might look great to the execs, but it doesn't mean anything to the average user." It's a lost opportunity if you don't share your story.

Unfortunately, many companies miss out on a potential connection with people who visit their sites by not having their unique story there. Everybody loves a good story. It doesn't need to be a book, just an entertaining and authentic story. There's a huge difference in connecting with a company if the choice was between one that just listed products or the other that started in a garage, and they were confident they could build a better mousetrap and went all-in. You want people to relate to you . . . so relate to them.

Customers would also like to know if you're socially responsible or if you have a passion for a particular non-profit cause. Including some personal information about your team members can also create a connection. Do you love sailing, red wine, or hiking? Let your customers know. These tidbits of information can go far in getting new customers and creating loyal customers.

Your Company brand is different than your product brand

As a marketer, it bothers me to see (and I see it more frequently than I should) a company confusing or mixing their identity with a product brand. If your company has three product brands, you need four stories—one for your company and one for each brand. Your company brand needs to be the big umbrella that shelters each product brand.

Proctor & Gamble is an American multinational consumer goods company headquartered in Cincinnati, Ohio. They either had or still have products under them such as Folgers Coffee, Pantene, Banker's Trust, Old Spice, Gillette, Oral B, Pringles and Kellogg's, to name just a few. Each of these well-known businesses have their own brands independent from Proctor & Gamble. P&G has it's own identity.

We work with a client who only promoted the brands of the company yet never the company. Although the products were selling, no one in the industry knew the company name. When they came out with a new product, it was always hard to penetrate the market because of lack of awareness. It has taken us several years to change this company's impression in the marketplace, and now when people mention it, the company's brand comes to their minds. Today, when this company brings out a new product, the streets are paved for it because people recognize the company as a powerhouse.

As you are seeking out your identity, make sure to differentiate it from your products' identities.

Why be a Jack of All Trades When You Can Be a Master of One?

Nobody wants to work with a generalist. When I get sick, I go to my family physician. However, if I had my druthers, and I had a searing pain inside my head, I'd rather see a neurologist. If I had severe stomach issues, I'd prefer to go straight to a gastroenterologist. Wouldn't you? Those are the true experts for those conditions.

Many people are afraid of niching down, thinking they are leaving opportunity or money on the table. That couldn't be further from the truth today. There is too much information out there to spread yourself around too much, to try to be everything to every market. Sure, you get a wide market but whenever they investigate who you really are, they realize that you have no expertise, no depth. You are not an expert at any one thing and they only need that one thing.

Securing yourself in a niche differentiates you. The more specialized you can be, the more credible you are. Let's suppose you met two lawyers, one did Immigration Law, Corporate Law, Labor and Employment Law, Personal Injury Law, and Trusts and Estates. The other works exclusively with men who are getting divorced.

Which one, on paper, looks as if he just might be the best at what he does? Two weeks after meeting them, you may forget about ever meeting the Jack-of-All-Trades lawyer and all that he does, but you'll most likely remember the specialist. If a male friend of yours was about to get divorced, it would be likely that you would refer him to that lawyer for two reasons.

1) You remembered him; he stood out.

2) Even though you only met him that one time, you believe that he is an expert at handling divorces for men because that's all he does.

When seeking to discover your identity, focus on what your company is really, and I mean really, good at. The market is looking for companies to quickly, yet expertly, provide insight or the answer (via a service or product) for their problem.

What makes your Company Special?

What do you do differently than anyone else?

- Do you have a specific methodology you use to create new technology products?

- Does your quality assurance program have steps in it that your competitors don't bother with?

- Do your key integration partners contribute to your new product development?

- Do you track customer service issues so that you can resolve future issues in your product updates?

Your unique selling proposition (USP) creates a perception that separates you from your market competitors. Your USP doesn't need to be about your product – it can be based on development processes, customer service or other value-adds you provide your customers. Your USP starts with a strong logo and is reinforced by your customer niche or specialty, a consistent image and a strong tag-line to communicate that unique selling proposition in all your marketing communications.

The Benefits of Having a Consistent Corporate Identity

A good brand is a powerful asset. It distinguishes you from competitors. It helps you carve out a "share of mind." Your brand makes it possible for you to compete on factors other than price. And when you're the "first call" solution for the product or service you sell, well, that's branding at work.

Once a strong logo is developed, you can create a consistent brand by adopting brand standards—a set of design rules that tie together the look and feel of all the marketing materials. Everything your customer sees—your website, your catalog, your brochure, your business card, your digital marketing assets, even your invoices— either reinforces your brand or diminishes it. When developing a brand standard it should include items like placement and sizing when using your logo, define the basic graphic elements to keep your brand memorable, and describe the color palette and fonts to be used.

Branding doesn't happen overnight. Branding is the sum total of every interaction your company has with your market. It helps if you have highly satisfied customers who spread the word about superior customer service that keeps people coming back. Targeted and plentiful advertising that entrenches your company name in potential customers' minds. A public relations campaign that leverages trusted media sources to establish you and your company as experts and positive forces within your tech community. Your brand comprises all of these things and cascades to all materials.

We all like to do business with people and companies we feel we know, like, and trust. If that's true, which it is, it would stand to reason to find ways for consumers to feel they know you, like you, and can trust you. Understanding your identity is the origin of marketing your company correctly. Identifying your corporate story and sharing it through the appropriate channels gives consumers the opportunity to feel they know you. At our core we are empathetic beings; we don't have to go through what other people have gone through to understand what they went through. Allowing people to empathize with your company helps build brand loyalty.

A company that invests in establishing their identity sends a message that you are a serious player in the marketplace and are intent on being successful. It shows that you take the time to work on the details early on and it gives your customers a sense of trust.

The best thing you can do, whether you are about to start a company or if your company has been around for decades, is to figure out your identity. You need to know your story, your strengths, what problems you provide solutions to, why your product/service is needed, why customers should buy your products instead of a competitor's products. Once you sort that out, you can build on that and always stay true to your mission and vision.

Naming the Product or Service

What's in a name? The answer is "everything." Company, brand and product naming should fit three criteria—strategic, linguistic, and legal.

Strategic criteria:

Does it capture your brand's essence in a meaningful way?

Is it appropriate and appealing to your target audience?

Is it as brief as possible?

Does the name connect to what the business is about?

Does it have the potential to be memorable?

Does it limit you in any way?

Linguistic criteria:

Is it appropriate in meaning in all major languages?

Is it easy to spell and say?

Does it limit you in any way?

Have you considered all relevant cultural sensitivities?

Is it too similar to an existing trademarked brand that it may cause consumer confusion?

Legal criteria:

Can you use it without infringing on another trademark?

Can you own and protect it as your trademark?

Is the domain name available?

Can you use it and protect it in all relevant geographies?

The naming process also comprises a thorough understanding of the target market and competitive landscape. The final name needs to be memorable, filled with meaning and sounds good when you say it out loud.

Keeping all of the above in mind, we use a four-step process to develop names: Discovery; Ideation; Proofing, Edit and Naming; and Tagline Creation.

Step 1. Discovery

The discovery stage includes review of materials, meetings with the team and research to dive into your brand and product strategy.

Step 2. Ideation

Using the naming considerations described above, start discussing and gathering ideas and a list of potential names. Develop a list of deep and broad potential names.

Don't copy a competitor or use something that's not legally available. Perform an initial internet search to verify usability. It's also wise to search the US Patent and Trademark Office website for previous uses. It will save you money and angst if your selected name is already in use.

Step 3. Proofing, Edits and Naming

Once you've had an opportunity to review the initial list of names, have team discussions to discuss the potential candidates, and perhaps work out more names based on that input. After a few rounds of discussions and additional input, you should arrive at the final name. The final name should stand out from its competitors and represent a new era for the company.

Once that name is selected, I highly recommend applying for a trademark from the US Patent and Trademark Office to protect it. If your sales will be international, also consider additional trademarks.

Step 4. Tagline Creation

This process is similar to the naming process. Through discussions and input from your team, create a variety of taglines to give the product clarification of use, market, model or other needs. Taglines should also be reviewed for basic trademark use also.

Logo Design

The first impression of your company, and ultimately your product, is your company logo. It should be professionally designed and it should be able to represent you well regardless of the application. For instance, it will appear small on your business cards and large your building signage. You may want to include it on t-shirts and hats or on your company vehicles. A good design will adapt to all these uses and more.

Potential customers will infer a substandard product from poorly designed materials - this includes your logo. This is one of those areas where you just shouldn't do it yourself.

But sadly, we once had a client who did just that. The company didn't want to invest in creating something professional looking and hired their neighbor's high school kid, Ken, to create it. It was so poorly designed that it made it impossible for us to create anything with that logo that could be taken seriously. This product was never successfully launched. Internally, we still refer to those homemade jewels as Hein-E-Kens - the "designer" Ken's signatory on his email. The one thing we do admit about this strategy: it was memorable.

The takeaway is design is never neutral. Graphic design can either support your message or detract from it. Design interrelates with everything you do. Whether it's a brochure, a catalog, a website or your packaging, you want everything your company produces to carry impact.

Color

Color selection is a big consideration in developing your logo. Colors really do have meanings. Would you consider working with a bank, a financial institution, that displayed a pink and purple logo? Probably not, unless you are a six-year-old girl.

Consider your industry and your company culture when making color selection. The link below shows meanings of colors. It's a good guide to choosing the right color palette for your needs. Color selection infographic: http://bit.ly/colors-meaning

Font

The font, or letter style, is probably the next most important element of your logo. The first consideration is the demographic of your audience. Is it a senior citizen who can more easily read a larger, serif font? Younger audiences would gravitate to edgier fonts - especially with a technology product. Kids will react positively to fun, playful, and childlike fonts. This being said, a competent designer or agency will steer you to selections appropriate for the audience you are striving to reach.

Typeface

Typeface refers to how text is arranged on a page or document. The typeface used, whether it is bold, in italics, colored, underlined, large or small, and issues such as line height/spacing, can enhance a theme, add personality, reinforce an idea, or demonstrate an emotion. It can draw you in and grab your attention. It can even increase trust in the product. And the point of all of it is to keep people reading.

Packaging the Product

B2B and consumer-focused packaging shows product offerings and supporting information to potential customers. Your philosophy and vision for the business will be some of the key content that helps drive the packaging strategy. If your focus is 'green', it may make sense to use minimal packaging with recycled materials. For a luxury product, the packaging needs to be very upscale and unique. It also needs to clearly exude a sense of luxury to potential consumers.

Color schemes, images, photos, logo elements, fonts, copy and all other aspects of your product packaging are an extension of your brand and must accurately convey the vision, mission, direction and level of quality that your organization and products represents. Research has repeatedly proven that consumers may infer substandard impressions of products from poorly conceived marketing and advertising material - packaging is no exception.

Your packaging design should include all appropriate content including product name, description, color or version name/description, cross-product usage recommendations, legal content and warnings, manufacturing information, bar codes, and any additional information you desire or is required based on the specific product.

Once the initial package design is created, be sure to extend that look to the family of products as appropriate. The design elements of the packaging must be consistent with the brand look and feel.

Design by Committee

Most startups and many established companies realize the

advantage that a great culture provides to the work group and the working environment. They truly work as a team. There are shared activities – and many times shared decisions. I greatly encourage this behavior and practice it at Marketing Matters. The one thing I won't do by team (or committee) is design approval. I highly recommend you don't either. Here's why.

Design by committee always takes two to five times longer to get consensus and it's virtually always a compromised logo, brochure or whatever it is you are designing. The team concept of agreement and buy in can still work though. First, assign one person to be the final approver. It's typically your VP of Marketing or the CEO in a smaller company – someone that understands the basics of design, works well with your design team and sees the company vision. Let the rest of the team know who is responsible for final approval. When you are presented design concepts, work with your design team to streamline and narrow down the concepts to the most appealing design or designs - the finalists. Once you get to that stage, ask your team for input to help you make your final decision. Believe me, this process will save you a ton of time and no one will feel slighted by it. This will also keep your design team sane. There is nothing worse than sending logo or brochure concepts and having four or five people give input on changes at different times. Your 33 changes can be done yet it become a painful, expensive, unnecessary waste of time for everyone.

Chapter 3

Know Thyself

"Know Thyself"

Plato wrote The Phaedrus about 370 BC. In it, Socrates tells Phaedrus that he has no time to ponder on mythology and other matters. Socrates says, "But I have no leisure for them at all; and the reason, my friend, is this: I am not yet able, as the Delphic inscription has it, to know myself; so it seems to me ridiculous, when I do not yet know that, to investigate irrelevant things."

Interestingly enough, the character Socrates got the credit for saying "Know Thyself," even though he actually never said it. The message has just been condensed throughout the years. Yet the meaning of what he said is what jumps out to me. Basically, Socrates was saying, the first thing I need to do is to know myself. Why would

I spend time discovering anything else if I still haven't figured out who I am?

Strength in Numbers

If you've gone through the steps I mentioned in chapter 2, you should have an acute sense as to what your company is all about. The next step is to get involved in the right tech communities. While there are a large number of tech communities available today, technology companies will typically fall within one or more of the three main organizations:

The Consumer Technology Association (CTA) (www.cta.tech) is a proponent of innovation and advocates for the entrepreneurs, technologists, and innovators who mold the future of the consumer technology industry. CTA provides a platform that unites technology leaders to connect and collaborate, and it avidly supports members who push the boundaries to propel consumer technology forward. CTA provides national and state advocacy on behalf of its members. Their research is top-notch and a free benefit to members.

CTA hosts the International Consumer Electronics Show each January in Las Vegas — 50 of them thus far. They are also very involved in creating technology standards that help provide some level of consistency for both manufacturers and the consumers who purchase their products.

CEDIA (www.cedia.net) is the global authority in the $14 billion home technology industry. CEDIA represents 3,700 member companies worldwide and serves more than 30,000 professionals who manufacture, design, and integrate goods and services for the connected home. CEDIA represents the "installed" home technology. The CEDIA channel is very cohesive and, to do business successfully in the channel, it's imperative to be a member of this community.

InfoComm (www.infocomm.org) is the oldest technology association. Established in 1939, InfoComm is the trade association that represents the professional audio visual and information communications industries worldwide. Members include manufacturers, systems integrators, dealers, distributors, independent consultants, programmers, rental and staging companies, end users, and multimedia professionals from more than 80 countries. InfoComm provides education, AV standards, thought leadership, research, and events, including international trade shows.

However, depending upon your specific product or service, there are even more organizations even more specific to your vertical channel, such as:

TecHome Builder (www.techomebuilder.com), which provides resources for builders who incorporate technology into the homes they build and to manufacturers that want to market to these builders.

Security Industry Association (www.securityindustry.org) members have access to education and training, certifications, government relations, standards, technology, and events specific to security.

HDMI (www.hdmi.org) The High-Definition Multimedia Interface founders joined together to define a next generation digital interface specification for consumer electronics products. The HDMI spec is also supported by major motion picture producers, satellite, and cable companies. They offer product certification to manufacturers.

DPL Labs (www.dpllabs.com) Jeff Boccaccio at DPL Labs is one of the industry's foremost experts on high-bandwidth digital signaling solutions. DPL offers the Digital Performance Level Program (DPL Program), which is a digital HD product testing program offering "DPL Seal of Approval" for those products that pass their rigorous testing protocol.

Imaging Science Foundation (www.imagingscience.com) was founded in 1994 by Joel Silver. ISF is in the display standard industry and is dedicated to improving the quality of electronic imaging. Joel and his team consult with manufacturers regarding product development, provide certification training, work with the media to arrange for test equipment availability for qualified reviewers, and offer ISF licensing based on standards for video performance or products that can be calibrated for specific brands optimal performance.

Wireless Speaker & Audio (WiSA) Association (www. wisaassociation.org) WiSA was established to foster interoperability compliance testing between consumer electronics devices and high-performance wireless speakers, and to promote the adoption of WiSA technology worldwide.

This is only a sample of the tech communities available today. There are many more associations available, with additional specificity, where your product category may fall into.

Chapter 4

⚜

Budgets and Marketing Strategies

In the previous chapters I wrote about:

- The importance of telling your company story

- Building the strength of your company brand and including it with your product brands

- Showcasing what makes you special—what really differentiates you from competitors

- The credibility that comes with creating a niche

- And, why you need to have a consistent corporate identity

Those things are fun to talk about because they are 'big picture' ideologies. Most business owners love to think conceptually. It's a gateway for them to daydream of future success. However, concepts are like shadows. Just as a shadow of a lion can't bite you, thinking conceptually won't make your business grow. You need substance.

The Identify stage is where you begin to develop your budget, marketing messages, strategy, and your marketing plan. It's also where you identify your sales channels, i.e. developing partnerships for OEM opportunities, with manufacturer representatives, distributors, B2B partners, or consumer sales. If you're like me, you like to start with the money. So, let's talk about how to develop a budget.

Develop a Budget

When we speak with potential clients, one of our questions is, "Do you have a budget?" For some reason, many people/companies are reluctant to tell an agency what their budget is. I believe their thought process is, if I tell them the budget, they'll spend all the money and I won't get the best return on my investment.

Unfortunately, the reason they feel that way is because it's either happened to them or to somebody they know, so it's not an entirely crazy manner of thinking. The fact of the matter is that it's not in anybody's best interest—the client or the agency—to put together a plan that zeros the marketing budget. At times, it's particularly frustrating for me to work with clients like that because I'm not great at reading minds. Not only does a budget need to be developed in-house, but the relevant parts also need to be shared with the agency so you can optimize the agency's work to ensure you get the most bang for your buck.

Marketing Budget

A marketing budget is exactly that, figuring out how much money you can invest in your company through the different forms of marketing. If your business is a flower, the marketing budget is the container that holds the water to be poured on the soil around you. Many companies that don't survive are typically the ones that don't have a marketing budget; they're like flowers dependent on occasional rain.

Employees have the luxury of waiting for decisions to be made for them. Often they will not ever know how much went in to making a company-wide decision or how long it took. Business owners can't sit and wait for decisions to be made, they need to make it happen. Once your business doors, virtual or physical, are open, decisions need to be made constantly. At times, you'll have a little time to mull things over and come to a well-thought out decision. Getting things done is more important than waiting until it's perfect. You can always adjust as new information arises.

Pre-planned marketing budgets can help guide your decision making in many different ways. For one, you can choose your priorities by figuring out what you have and what things cost. Here's an example:

Let's say you expect to have $150K to invest in marketing for the upcoming calendar year. Yet the three major trade shows you want to attend are going to cost you approximately $130K in total. What do you do? By knowing your budget, you can base decisions on what is going to be most effective for you. You may decide to go to two out of the three shows and reserve some resources to launch a content marketing program and social media to announce new products or new client programs. You could also set aside funds for a public relations campaign or invest in some digital marketing to amplify those launches.

Here's a sample of a marketing budget we use to help our clients: http://bit.ly/mktg-budget

In-House or Outsource Marketing

A prevailing thought is that hiring out is always more costly than hiring employees to handle all of the marketing in-house. This isn't necessarily true, especially when you consider the finished product or results. To understand the cost difference between having the marking done in-house as opposed to the benefits of hiring an agency, a lot of factors need to be weighed.

Hiring your own in-house marketing staff involves the ability to find and select the correct talent—people with the right experience, education or job experience, attitude, and industry-specific knowledge to make your company more successful.

The budget expense of not having the appropriate skill level and expertise can lead to turnover (voluntary or involuntary), repeating the hiring cycle, and other unanticipated costs. This might be why 50 percent of digital marketing activities, for example, are outsourced (Gartner 2013).

Other budgeting costs include time.

- How much time does it take to recruit, interview, and hire someone?

- Is your time lost from more productive tasks?

- Is the adequate amount of time being spent on bringing your team up to speed?

According to QuickBooks, employers pay the following costs for employees before benefits, perks, and before considering recruitment, turnover, etc. (Poladian, 2015).

Social Security Tax	6.2 percent on each employee's wages up to $118,500
Medicare Tax	1.45 percent of each employee's salary (plus 0.9% more if salary exceeds $200k)
Federal Unemployment Tax (FUTA)	6 percent of the first $7,000 of employee wages.
State Unemployment Tax (SUTA)	Tax rates and policies vary by state. Employers can take a credit for the state unemployment tax they pay against the FUTA tax. The maximum credit is currently 5.4 percent.

Workers' compensation	$1.85 per $100 of payroll, or 1.85 percent of an employee's salary. (higher in CA or high-risk jobs).
Payroll provider	Various fees

Beyond the basic taxes, Leslie Shiner of the Shiner Group also calculates the budgeting cost of an employee:

Let's look at a sample employee who is paid $25 per hour. If you add some sample percentages for payroll taxes, workers' compensation and liability insurance, you can see that it now costs you $34.50 per hour to pay that employee.

These costs are considered variable costs; they directly relate to your total gross payroll. If payroll goes up 10 percent, then each of the burdens above will also increase by 10 percent. But there are other types of labor burdens that are fixed costs; the costs are the same, no matter how much you pay your employees. These include such benefits as health insurance, as well as vacation and holidays.

For example, let's assume that the cost of health insurance is approximately $500 per month, or $125 per week. If an employee works 40 hours, then the burden of health insurance is $3.13 per hour ($125÷40 = $3.13). However, if your employee is paid hourly and only works 25 hours in one week, then the cost becomes $5.00 per hour ($125÷25 = $5.00). So, while the cost to you does not fluctuate, the cost per hour does.

Do you pay vacation and holidays? Consider the fact that 10 paid *holidays per year adds another 4 percent burden for your employees. For your $25-per-hour employee, that adds an additional $1.00 per hour to the burdened costs. Let's see how to calculate that number. The standard number of hours in any given year is 2,080 (eight hours per day times five days per week times 52 weeks per year.) If you pay for 10 holidays, or 80 hours, then 80÷2080 is 4 percent. Four percent of $25/hour is $1.00. If you provide two weeks paid vacation, that adds an additional $1.00 per hour. Looking at this sample employee and adding these additional burdens, it now costs you almost $45 per hour.*

Some other budgeting costs include:

- *Mobile phones, computers, office setup*

- *The time it takes a new employee to assimilate to the job and begin to contribute*

- *Leaves of Absence for maternity and paternity, illness, bereavement, legal costs, and other reasons*

The Shiner Group makes another great point in regard to cost efficiency, which relates to the time paid for nonproductive work; such as meetings, trainings and certifications, driving time, equipment, paid sick leave, pension and a myriad of other costs.

These costs will decrease the total utilization of each employee's time by an estimation of 30 to 75 percent. Shiner argues the employer does not recoup the loss of full utilization by billing clients, and that it absorbs those costs.

In short, don't believe the myth that hiring an employee (or employees) to handle all your marketing will save you money. The truth is that with an agency you will get truly proven professionals who overperform and deliver at a fixed price. You'll pretty much know what you're paying for if you outsource—pure marketing results.

Marketing Plan

Your marketing plan is an ongoing, living, dynamic document that serves as your guide and keeps you on target for achieving your goals and helps you identify the critical tasks for marketing your product. This dynamic document can (and should) be continuously reviewed to keep up with changing conditions throughout the life cycle of the plan.

The development of a plan is the cornerstone of marketing your product. Operating without a plan is a little like driving to Washington D.C. without a starting point, without directions, and without a time frame. You will spend a lot of time and money going nowhere.

Over the past two decades, we've worked with many start-ups that didn't have a plan. That's okay. That's why they brought us on board to help. We work with those folks to develop realistic goals for creating support for and onboarding new dealers while raising awareness for their products—and typically succeed.

I can virtually guarantee failure when I see companies who, on behalf of their funders or board, insist that their goal is to sign on 300 new dealers in three months as a start-up. That's unrealistic—

especially for a sophisticated product, like a control system. It takes about 11 'touches' to really be noticed. It also takes about seven hours of communication (live and digital) to form a strong enough bond to begin the reseller relationship.

Another path to failure is the "Squirrel" syndrome. Every shiny new opportunity that crosses their path (a new SEO campaign, adding unbudgeted trade shows, signing expensive consulting contracts without performance guarantees, etc.) takes these people off track from their goals and dilutes the already limited resources they have available. I'll be the first one to recommend a fine-tuning of the goals and plans based on data. One of our company mantras is test, measure, and adjust. It's another issue to chase after everything they come across.

You absolutely should start with a strategic marketing plan, however, you must also remain open to opportunities not necessarily in your initial plan and exercise foresight about upcoming trends and industry changes. Failure to do so can lead to disastrous results. Note the case of Blockbuster, which was not alert or open to changing their business or marketing plan when an unbelievable opportunity came along:

Blockbuster was approached multiple times throughout the early 2000s to purchase Netflix for $50 million. Blockbuster refused. Instead, Blockbuster went into business with the doomed Enron organization, who had ventured into telecom and attempted to purchase Circuit City, even as it was declining. They failed to compete with Redbox and Netflix, which now has a market cap of $19.7 billion, and ultimately landed in bankruptcy in 2010. They ended up getting bought out themselves, by Dish, who also later regretted the decision and shut down Blockbuster for good in 2014.

If Blockbuster had read the trends correctly and allowed themselves to deviate from their original plan, they would have seen that streaming was going to be the wave of the future and had they bought out Netflix, they would have been more profitable now than ever.

Marketing Strategy

Your marketing strategy basically aggregates all your marketing goals into a single plan. Ideally, your marketing strategy is based on research and/or experience. That research doesn't need to be formal. From my experiences over the past 20 years, not many start-ups (or even established companies) are in a position to pay for expensive research. There are other ways to get the information you need though.

The Consumer Technology Association (CTA) has an incredible amount of research available to its members at no charge. You can get stats on topics from high-resolution audio adoption, the path to purchase using mobile devices, to consumer sentiments on augmented reality and virtual reality and the state of the building industry with technology. In addition, CTA has developed a plethora of standards for audio, video, television, DTV interfaces, portable handheld and in-vehicle, health and fitness technology, consumer electronics networking, modular communication interface for energy management and residential systems. You can access that research and the standards at https://cta.tech/Research-Standards.aspx.

For the pro AV and information communications industry,

InfoComm also offers a variety of research and standards. On the residential technology side, CEDIA is your go-to source. There is also a ton of content developed by end users. Use Google and find out what they have to say about your product category. There are a ton of tech forums that likely have some of the answers you are looking for.

Some other tools to use to develop your strategy include collaborating with others, including influencers, who are part of your target market. A dealer summit that brings your top customers together to review what worked well and provide input on new product concepts will help give you the guidance you need to develop the right products. Experiment with new channels and platforms (with a solid plan). Get your employees involved. They are top brand advocates and can give you insight on the company, products, and customer needs. Partner with synergistic companies. If you are an audio receiver manufacturer, it makes sense to partner with loudspeaker brands and cross-promote programming and other assets with your respective customer bases.

Don't be afraid to experiment and try some untraditional methods when developing your strategies. Pop-up stores, crowd funding and other non-traditional marketing techniques can be very effective in growing your brand.

Marketing Messages

Your marketing messages are the content you develop to differentiate your product from others on the market. These core messages will be used on your website, in advertising, in your product literature, in your email marketing campaigns, and will be included in your trade show displays.

First, your message needs to focus on solutions, not features. When I worked at Niles Audio in the 1990's, we launched a powerful campaign for the new outdoor speaker line. One of the main benefits the OS speakers provided was tremendous bass output when placed in a corner under a soffit. We came up with the headline, "Corners Beautifully," with a close-up angle shot of the speaker for the ad campaign (remember – this is pre-Internet). It was highly successful and very memorable.

For an email campaign, a headline and subtitle can work very well to get readers' attention. Use unexpected, thought-provoking titles to make folks want to read more and increase your open rates. People love lists. Titles like, "5 Mistakes Integrators Make with HDMI" or "7 Secrets to Network Extenders" will be well read articles.

Offer a gift or marketing tool. Use a small gift—a whitepaper, tips or cheat sheet, an online tool or some other relevant giveaway to get your readers to learn more about your products and how they solve the problems your audience is dealing with. These have value to an interested prospect and will increase your response rates.

If you are considering a direct mail campaign, a lumpy mailer works very well. What is the first thing you open in your mail? That lumpy thing (right after the checks). An intriguing message on the outer envelope with compelling content will drive your prospect one step closer to making the sale.

The bottom line is you need to use messages that resonate with your prospect. Think "What's in it for your customer" when developing your messaging.

Sales Channels

When I talk to start-ups, I'll frequently ask them, "How do you plan to take your product to market?" And in many instances, they'll reply, "Gosh, I don't know." So, I'll offer them an overview of the ways to bring products to market. The type of sales force you should use largely depends upon the type of product you have.

Direct to Consumer

Depending upon your product (i.e., how it is used, whether it requires installation, whether it will require updates, or its complexity) this may be a method to employ. Many start-ups want to go consumer direct because they feel this is an easier path to obtain funding. But I can tell you, it's not as easy as you may think. It takes a fairly hefty budget to get on the radar of consumers, who are bombarded with offerings throughout every minute of their lives. Consumers have been conditioned to tune out much of this advertising, rendering it ineffective.

A consumer is less likely to read a lengthy blog post than just five years ago in our new ADD society. Using tools like an interactive quiz or collaborating with an influencer will take more effort, and potentially more budget, to develop yet with the right content, it will be more effective.

Manufacturer Reps—Paid a commission for the sale

Manufacturer reps typically work with integrators or retailers within a specified territory. They never take possession of the product; they're paid a commission on the sales that are done within their territory or with specific customers per your agreement with them.

In many instances, the reps are working with a full portfolio of products. Although they'll sign up to sell your product, they may not take the time required to make hiring them lucrative for you. They need to clearly understand the benefits of your product and have relationships with dealers or retails who use or sell those products. You also have to provide the right support and materials for them to sell your products.

Distributors—Purchase the product from you, keep it in inventory and resell to dealers or retailers

There are national and regional distributors. Your reach and support, as well as the product you are selling, will determine which distributor(s) would work best for you. For example, if you know you aren't ready to service the entire country, go with regional distributors. Even if you go with national distributors, it is not uncommon to use some regional distributors to fill in some gaps in coverage.

Like manufacturer reps, if you decide to sell through distribution you will need to give them the materials and training so that they can easily sell your products. Having said that, the product needs to be a bit simpler to understand or something that, say a custom integrator, is used to working with. The killer here can be the learning curve.

Another consideration with using a distributer is that they're going to require marketing financial support, not just marketing materials that perhaps would be sufficient for manufacturer reps. They may ask you to participate in different training events or shows that they host and ask for marketing funds to participate. In most instances this is a very effective way to get in front of your target audience.

The main thing to keep in mind when working with distributors is your margins. You need to give the distributor sufficient margin for him or her to make money between what they pay you for the product and what they sell it for. So, the caveat is that you won't make as much for each sale but the benefit is that you'll have consistent sales using their sales teams.

Retail—Purchase products directly from you to sell to consumers

Retailers are yet another option. You need to be sure that your product can sell itself based on its packaging and point of purchase (POP) displays. Getting into the retail market has its own sets of challenges as the market is flooded with a different assortment of products. Without a track record to boast, retailers may hesitate to bring your product on board. Again, packaging and your POP display is key. You better have a really good four-color box that explains well what problem your product resolves and how it does it. Visual aids are very important in the packaging as well in your POP displays. Focusing on features is good, yet the benefits to the end-user are what's really important.

Direct Sales Force—You hire salespeople to sell your products to integrators, retailers and/or consumers

You can hire a number of folks who work with your target

audience, whether they're consumers or through the B-to-B channel. This can be one of the more expensive ways to go to market if they are direct employees due to the overhead. Earlier in this chapter I did a breakdown of costs for employees. Those same costs are also a factor if you want to hire a direct sales force. One of the biggest costs becomes travel and lodging, something you don't have to pay for if you use manufacturer reps or distributors. Another option for a direct sales force is to hire based on commission. These are typically independent folks who will work with manufacturers much like a manufacturers rep. I'm seeing more of this in our current collaboration economy.

Even if you work with reps and distributors, you will need a strong sales person to lead and manage your outside sales force. Be sure to hire someone who understands the industry, has a proven track record, and isn't afraid to spend some time on the road to make those distributor and manufacturer rep relationships successful.

These go-to-market strategies are not exclusive. I'm starting to see some new hybrid methods that are effective for some manufacturers, especially those with complex products or that have products in new categories. For instance, one of our clients has a high-voltage power management product with a base price of $80,000 (before shipping or installation). They hired independent manufacturer reps to manage consumer sales rather than asking the integrator to make the sale (most integrators typically only deal with low voltage products). This takes a huge burden off the integrator, yet they still make their margin on the product sales. The close rates are pretty impressive. The reps know the product intimately and understand the sophisticated installation requirements and the type of customer that the product appeals to. In summary, you need to look at how complex your product is and then determine what's going to be the most effective method for you to take it to market.

PART 2:

SIMPLIFY

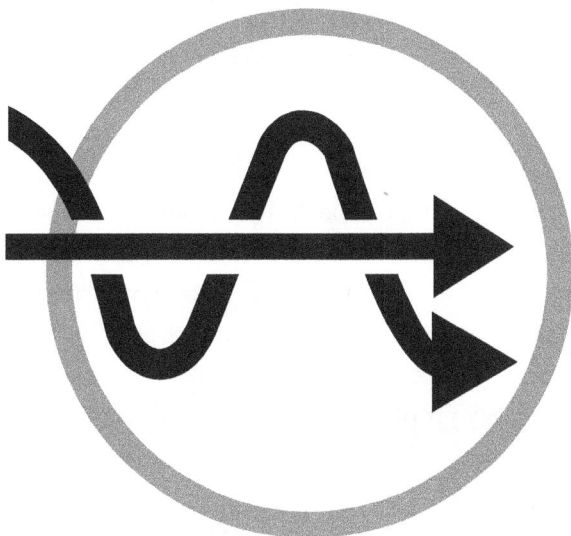

Chapter 5

Simplify

The world of technology, at times, seemingly moves at the speed of light. Take mobile phones for example. It appears that every other month a new phone comes out with features that trump the one that came out just two months prior. You can essentially handle most of your business on your cellular phone, shoot and edit video on your phone, shop from your phone, bank from your phone, and even keep your phone in your pocket when you go for a swim!

Mobile phone companies understand that the general public has been using mobile phones for decades now. In fact, it was Sept. 21, 1983 when Motorola made history with the first FCC-approved 8000X, the world's first commercial portable cell phone. Yes, mobile phones have been out for more than 30 years now! What's interesting about the way they are marketed is that they only need to highlight the new features. They don't waste time showing us things we already know they do, they show us the exciting new things they do.

Don't be so Technical

I believe most of the technical world can learn from how these powerful, global businesses market their phones. In their commercials, they don't show you how to use new features, nor do they show you the technical science behind how their new features work. They simplify things by showing us the benefits of how the new features make things easier for us. It's genius really. They don't have scientists with wire-rimmed glasses and pristine-white lab coats pontificate technical jargon because no one would understand it. Yet, to the detriment of many technical companies, technical jargon escapes from the lab-room/ production-room floor and floods their marketing.

Too many tech companies today tout the technical specs of their products instead of relaying the benefits of their products in everyday, simple language. You can have the best product in the world, but if the marketplace doesn't understand the problem it solves or the solution it provides, no one will buy it. You need to know the specs and how things work and your target audience just needs to understand the benefits of your product.

If you are a television manufacturer and use 4K Ultra HD to describe the resolution, your audience will understand that the display has great clarity. Using "3840 pixels x 2160 lines, 8.3 megapixels, aspect ratio 16:9" as your product feature won't communicate that message clearly to all of your audience. The point is, you want to communicate that this product has the highest resolution available to provide the ultimate user experience. The brilliant tech gurus are critical for successful product development. Some companies just need to take their product messages and translate it into simple features that the integrator and end user will understand.

The product development folks are just so familiar with their products that, at times, they don't realize when they are talking over the heads of those they want to sell the product to. Then they scratch their heads, wondering why it's not flying off the shelves. They make another mistake when they talk to their integrators about the product. They assume that the integrators are "sales people," and they will understand how to best position their product. Wrong! Just as the end-user needs to hear about the features that solve a problem or present a solution, the integrator needs to hear about specific benefits that will be appealing to how they sell and install the product. A smart home control systems may include every feature a consumer wants at the right price point, yet if the product benefits aren't clearly stated and the integrators' sales process isn't supported it will likely not reach a high level of success.

Integrators typically carry a few different lines of product for an installation need. If they can't clearly and easily understand the value proposition of a product, they're going to present another product, one they can better understand, or better said, one they can make their customer understand more easily.

Remember, at the end of the day people only want to know the reason for your product, not the inner-workings of it. Consumers don't buy "black boxes"—they buy the experience and benefits it will provide.

This is important: you need to have a clear message that simplifies what your product does. That message needs to be written in every marketing avenue—website, brochure, digital marketing, etc. —and it needs to be articulated by your sales and marketing people. Being consistent and easily relatable is the best way your message can spread across the country and around the world.

Policies

Now that I mentioned a consistent message, I want to talk about its importance. Many startup companies grow too fast and the business owner either doesn't have the time or doesn't know that he or she should develop processes or procedures for the employees to follow, otherwise known as policies.

It's mindboggling to me how some companies run departments that interface with their customers; the two most prevalent of course are the sales and customer service departments. In today's world of shared information and experiences, particularly through social media, one would think that a business owner would take the time to have procedures in place when a customer is dissatisfied, for whatever reason.

The lucky ones happen to hire people that are naturally empathetic pleasers, however, without understanding how to resolve an issue, even people perfectly suited for customer service roles can't fully resolve an issue.

For some reason, many tech companies think they don't have to have people available to speak with. We worked with a CRM software company that I believe does an awful job at customer service. This particular CRM has a chat feature for their clients to use on their websites so that visitors can quickly get their questions answered and problems solved. We had a problem with the CRM - the language on the chat option suddenly changed from English to Czech. That's a big problem. An even bigger problem is that this company doesn't have any Czech clients.

The only way of getting in touch with them was via email and their online support portal. After several days of sending them emails, they finally answered with, "The chat function should be in the same language as the website." And that was it. That was all they said. Needless to say, I was extremely frustrated with their level of customer service. We responded back and had to wait another three days for them to fix the language feature. After this experience, I no longer recommend this product to our clients.

Sadly, the technical space is littered with poor customer service like this. It's as if the owners just think about the production and sales but don't give a second of thought to servicing their customers.

Every company should write down or create videos with step-by-step procedures on how to solve their customer's issues. It's Business 101.

We have a particular client who utilizes our public relations and social media management services consistently, but who puts us in a tough spot because they lack a strong customer service process. They have a company phone number and use an email support portal for their customer service department. The problem is they don't always answer the phone or return calls or emails when someone reaches out. We promote them expertly and continue to help them acquire new clients but when they fail at the customer service portion their customers don't convert to raving fans of their products, and this is a missed opportunity for them.

Everyone talks a good game about how customer service is a high priority, but as the saying goes, the proof is in the pudding. The proof is if there are actual policies and procedures that anyone can follow to provide exceptional service to their customers. In our particular case with our client who doesn't always answer the phone or respond to emails, we do what we can to handle negative social media posts and resolve their customers' issues, but we will always be limited in our effectiveness if they never reply to their customers' complaints.

I've never understood why people wouldn't focus on customer service. The truth is that it takes a whole lot more effort and cost to obtain a new customer than it does to retain a current customer. By providing poor customer service; by not having policies in place for contingencies, you are shouting a message to your customer base that you only care about your sales and not about them. There are many reasons to take care to resolve your client's issues promptly and efficiently. These are my top two:

1. Client acquisition is always more expensive than taking care of your client needs.

2. You hurt your reputation, your brand.

At the end of the day, it's your job to make it simpler for consumers to do business with you. It doesn't matter if your selling to retail, integrators, or through independent reps or distributors;

over complicating your products messages and making it difficult for your customer base to find a resolution with an issue regarding your product will slowly kill your business.

In summary, some considerations in developing your simplified process include:

- Develop clear benefits for each of your sales channels. Why should they commit to your product instead of other market options? What part of the market (high-end, mid-market, mass market, etc.) is your product best suited for? How will the product benefit their customers? What profit margin is available for them?

- Make product training simple and easily accessible. Online training and one minute YouTube videos can go a long way in solving an integrator's consumer's issue during an installation or setup.

- Make sure you have set up tech support that is easy to reach using a variety of tools—chat, online, email and telephone access. An installer may have an issue that you feel is simple to resolve, yet if he can't reach tech support quickly, it remains unresolved and challenging.

- Make product registration simple. If it can be automated, do it! Product registration is a strong tool to track and resolve warranty issues and to use as a post-sale marketing tool. When you develop the next generation of the product or a must-have accessory, your customers will want to know about it.

- Build a robust customer service department with solid procedures to create raving fans of your customers.

PART 3:

TOOLIFY

Chapter 6

Online Tools

Tools have been necessary since the earliest days of humankind. Some say the use of our hands to create the first tools, as rudimentary as they were, is what set us apart from all other living creatures on this planet. If we look back a thousand years we can see the massive cities, enormous monuments, and other jaw-dropping marvels we have been able to create because we learned how to utilize tools. A century ago, people could identify a doctor as he would knock on the door making a house call by the size of the satchel required to carry his instruments. Today, many verticals are still identifiable by a person's tools. A worker can leave his tools behind and anyone who comes upon them can decipher if the person is a painter, plumber, or farmer. Tools have been, and always will be vital to the advancement of our species and, for the sake of this chapter, as us as business people.

To start up a business or market products in the 21st century requires the optimal usage of the tools available to us. In earlier chapters, you read about putting your marketing plan together, identifying and simplifying your message. Now let's get into setting up the backbone of how you're going to toolify (prepare the tools) to amplify your message. One of the most important tools to have is a website.

Your Website

A website is a must have. It is absolutely imperative, essential, required, vital, and every other synonym for the word needed. You don't exist on the Internet without one. I'll take that a step further, you don't even have a real business without one, especially if you're a tech company. In many instances, and for some companies in all instances, it's your image. We've all seen many horrible, clunky, wordy, abandoned, websites and we've seen very clean, efficient, updated, ones. Which would you guess are income-generating assets? The question is, how do you build an effective website today? I'll show you.

The look and feel of your website needs to be consistent with your logo and overall color scheme for all your marketing materials. Just by staying cohesive with all the other marketing assets you will give off a feeling of trust to your customers. If you deviate from the pattern you've already established, you may find that you may have to resell customers that already trust you because they will think this is something different or new for you.

Besides the color scheme/look and feel of the website, your home page needs to tell visitors who you are and what you do. Too many companies appear to insinuate that because someone is on their site they must know who they are. That's simply not the reality. Not only that, but it's also quite rude. I'll explain. When you meet someone in person, what do you do? What do you say? Most likely you extend a hand, offer up a smile, say hello, and say your name. When someone comes into your home or office, what do you do? You give them a welcome, right? Yet, online, many tech companies get right to business without a welcome, without a handshake, a smile, or letting the customer know who they are and what they do.

People do business with people. Sure, it's a website but you are interacting with people. Make sure you have a nice welcome there. It doesn't have to be the size of a blog, yet it should be clear to the visitor that you have taken into account that he or she feels welcomed to the site by a quick introduction.

Be prepared to invest time and money in your website. As I stated, in many instances it will be the only interaction your customers will have with you. Some businesses are set up in a manner that the people who work there, including the owner, will most likely never meet a customer in person. I'm speaking from experience because I have had customers that interfaced with me online for months prior to us ever meeting in person. I actually have a few clients that I have yet to meet in person, every bit of interaction has been online or over the phone.

A welcome video is a great tool to have on a home page. It doesn't have to be very long or professionally edited and produced. A simple one-minute video by the CEO or a high level executive who looks comfortable in front of a camera can deliver a warm welcome and let any visitor know what the company does, what they stand for, and the benefits of its product(s). Make sure it's someone that is customer-facing. We all like to put a face with a name.

A search functionality is also indicative to your visitors that you've put a lot of time and effort in ensuring they can have a great experience on your site. There are some companies that literally have hundreds of products listed on their site, yet it's not built with a robust search feature. If you have a lot of products, a product selection wizard could be a very helpful tool for your customers.

A free gift is also an indicator to your visitors that you appreciate the fact that they have found your site. The key words there are free gift. Don't ask for anything. Perhaps you can offer a white paper, access to instructional videos, or even a short downloadable e-book titled "7 Tips to Blah, Blah, Blah." Be mindful you not give away things that will lengthen your sales process. If you use the free gift to solidify your expertise or to hammer home the benefits of your products, it can very well shorten the customers' buying process. Remember, the whole purpose of giving away a free gift is to get a paying customer. In my experience, to move a prospect into a customer typically takes seven hours of communication, 11 touch-points, and four platforms to make the conversion. So fill up your arsenal with videos, white papers, published literature, newsletters, e-books, and other sorts of giveaways.

If your goal is to capture an email or phone number from your visitors, offer a sign up to receive your newsletter as a first step. I highly recommend having your newsletter signup on the header or footer of every page. Then, of course, produce newsletters that they will find informative and entertaining.

Links to social media are also key elements to humanizing your company. People do business with people, so give your customers the opportunity to follow you on social media and learn more about the staff and the company. If your company is the slightest bit philanthropic or does any sort of outreach or help in the community, a link to the social media pages where your community involvement is featured is a good way for visitors to learn more about you and hear your story.

The "About Us" Page

I have reviewed website analytics for two decades and, after the home page, the two most viewed pages are the About Us page and the Contact Us page. That has certainly been the case for our website. Too many companies don't bother using these often-visited pages to their fullest potential.

The About Us page is the perfect place to get into more detail about the people who make the company tick. As I discussed in Identify, tell your story here. Every company has one. What was the genesis of the company? Was it intended to be a means of passive income, and due to the benefits of the products or services it took off? Why do you exist? What motivates you to get up in the morning?

Another idea is to highlight your customers in this section. Tell the story of how they got to you and why you are their go-to company now. A prospect may be able to better identify with your customers because he or she is in their same situation.

List your leadership, if not the entire staff, depending on your size. Put a face to the name with an image and if there's a change, update it on the website. Not only will the customers further identify that you are real people, your staff will also feel more secure in their roles for the organization.

If you have any press releases, announcements or media placements, list them on your About Us pages as well. Feature the awards you've won and include links to your media coverage or make it clickable to read the entire segment in your site. If you're a big company with many press releases you may want to dedicate a separate website page for them all. If you're a small company and your website is already up and running, it might be easier to put them in the About Us section than reworking your navigation.

The "Contact Us" Page

Business 101: Make sure your customers can contact you. You would think everyone knows that. Yet, this is another web page that doesn't get the attention it deserves by many companies.

The essentials for this page:

- Address—If you have a physical location, list it. It shows you're for real.

- A Phone Number—Ensure that it gets monitored, meaning the calls are forwarded to a person that can not only answer the call but also that can be of help to the caller.

- An Email Address—It's perfectly acceptable to have the email go to an info@ or sales@ email address. Like the phone number though, make sure it gets monitored. Also, I recommend that you set up an auto-responder that goes out to whoever sends your company an email. The sender should get it within a minute and it should thank them for their email and let them know when to expect to hear back from you. Then, do as you say!

A cool feature to have on all pages is a chat window where people can ask whatever questions they have and chat in real time with someone who can help. Some companies that use this feature don't seem to understand that it could be turned on and off. During their work hours, they chat away with prospects and clients and when they leave for the day, people are still trying to communicate with them and are getting frustrated because they are using a chat feature that appears to be abandoned or doesn't work. It is perfectly acceptable to list the hours the chat feature will be monitored or simply have it not appear when it can't be manned.

There are so many intricate things to do in order to have a great website. I could write another book entirely on just that if I listed them all. I could go on and on about how to develop a great looking and highly functioning website since we've built plenty and there are a lot of other features I didn't mention. The particular features will depend upon what your products are, if you are selling online, what digital assets you have, your customer service needs and a plethora of other variables. Please contact me and I will personally answer whatever questions you may have about your website needs.

Landing Pages

I don't recommend partaking in digital advertising/marketing without first developing the place you want to send your customers and prospects to. That place online is called a landing page.

The first thing to do with a landing page is to ensure that your call to action is above the fold. The term, above the fold, came from the days when newspapers ruled. Above the fold meant when the paper was folded and you looked at the front page everything you could see there—the name of the paper, a picture, and three or four of their top stories—was considered above the fold. Online, above the fold is the area of a screen that a visitor sees as the site populates. Don't make people scroll down to see your call to action.

Whatever you do online has to be mobile friendly, so bear that in mind. Also, remember that we all have some form of attention deficit. Some people may go to your landing page ready to purchase your product, don't give them a lot of marketing or salesy jargon that could talk them out of it. The good stuff has to be at the top.

You only need one call to action on your landing page. If a prospect clicked to learn more or to purchase a particular product, don't give them three more to choose from! It's like when you take a child to get some chocolate ice cream, if you know he likes chocolate, order him the chocolate ice cream. But if he walks up to the counter and sees all sorts of assorted colors of ice cream, he may never make a decision. All you need in each landing page is one call to action. It works.

Have a clear value proposition, a promise that your product is of value. It's the most important part of a landing page because people will only take action if they understand and want the benefit you are offering.

Here are some other quick tips on landing pages:

- Avoid distractions. It's a landing page, not a website. It doesn't need an entire navigation site or links. They work against you as distractions.

- Do include your logo. You want to always burn your brand/ logo into their subconscious. It's particularly helpful when you have multiple products.

- If you're going to use images, you only need one. Have a warm, bright, high-quality image at the top that conveys the emotion you want. Beware of cheesy over-used stock photos of men in suits shaking hands, multiracial workers smiling, or a pretty young woman wearing a headset as if she's expecting your call.

- Ensure that your copy is clear about the benefits of your product and that it is written to your target market. If your target market is a Technical Director for a company, a more scientific/technical copy would suffice. If your target market is young millennials, your copy should reflect their language.

- Don't use exclamation marks. And really don't use multiple exclamation marks after a sentence. It looks as if you're shouting and people relate that to cheap used car ads.

- Don't require too much information for the visitor to take the next step. Make it simple. If you want to capture their email and phone number, that's acceptable, but to add additional information such as last name, city, state, hobby, profession, or what model car they drive is overkill . . . and it can kill the sale.

- Make sure that it's optimized for mobile devices. Period.

Video

About ten years ago, fresh content was king in Google searches, which is partly why blogging skyrocketed. However, there's a new king in digital marketing, all hail King Video!

We've all heard this saying; A picture is worth a thousand words. It has been added onto. Now the saying is, If a picture is worth a thousand words, a video is worth millions. I happen to have my own version, it goes like this, If a picture is worth a thousand words, a product video could very well be worth a thousand sales!

Consider this, video now appears in 70 percent of the top 100 search result listings. The biggest search engine in the world is Google. And it isn't even close. According to eBizMBA, a website designed to provide the web's best resources to help grow an eBusiness, Google gets more searches than the next fourteen search engines (Bing, Yahoo, Ask, AOL) combined! That really shouldn't come as a shock, as more than likely, you use Google whenever you do a search. Now, take into account that Google owns YouTube . . . let that marinate in your head for a minute. The biggest search engine in the world is in cahoots with the largest video search engine in the world.

Speaking of YouTube, they have more than a billion users—

almost one third of all the people on the Internet. YouTube overall reaches more 18-49 year olds than any cable network in the United States. In fact, not to be caught behind the times, in March of 2016 the Television Academy announced it has expanded and redefined many short-form categories for this year's Prime Time Emmys. The new categories are:

- Outstanding short Form Series, Variety

- Outstanding Individual Actor in a Short Form Series

- Outstanding Individual Actress in a Short Form Series.

Yes, actors can now win a Prime Time Emmy on YouTube. Now, lets talk video for business purposes. According to Kissmetrics Analytics, people are 64-85 percent more inclined to buy a product after watching a product video. This is simply a marketing force you can't afford to ignore.

Remember, attention deficit is abundant in today's society. Most people drive with a steering wheel in one hand and their mobile phone in the other. It's our reality. Did you know that more than half of YouTube views come from mobile devices?

The question is no longer, should we use video? That should be a foregone conclusion. The real question is, how do we leverage the power of video to maximize exposure and sales? Here are some tips:

- Try to keep your videos under a minute. Any longer and you could lose your audience before they get to see your call to action. If your video is on several different topics so it would take more than a minute to say everything, shoot multiple videos.

- Your phone is an acceptable video recording device. Many people don't produce videos because they feel that the video should be exactly that, produced. That it should be perfect. Let me share with you something very important. Done beats perfect. If you wait for it to be perfect, it may never get done. Use your mobile device and just shoot your video. It's acceptable. In many instances, you will come across as more genuine and authentic if your video isn't perfect.

- Mind your visual and audio. Just because you can produce videos from your phone doesn't mean to ignore practical sense. Make sure you have proper lighting and that your message can be clearly heard.

- Make sure your head is at the top of the screen. A nice tight shot, from the waist up, so that your face is clearly seen, is ideal.

- If you have impactful visuals, use them. Use music or a voice-over and show the images you know will sway your viewers.

I strongly recommend you have videos for all your product overviews and installation tips.

Unboxing videos are an extremely effective tool. Those are typically videos that your customers create and post. The video typically starts and the customer or reviewer talking about your product, which is still sealed in its package/box. They open it and discuss the features and benefits. Some even go as far to demonstrate the product in action. I strongly recommend you give away a few of your next new products to a few of your favorite customers, in exchange for him or her to shoot a video of opening the package and authentically speaking about it. Then, use your marketing muscle to share it. Unboxing videos come across as a genuine testimonial and many times are more effective than the videos your company produces.

The number one thing to do when shooting a video for your company, whether it's a product video or a welcome video is this— provide a call to action. A link to a landing page is a good next step to ask your viewers to take. Make it easy for those watching your video to get in touch with you. Include your website addresss, an email contact and/or phone number at the end of your video or in a footer throughout the video.

A balanced approach is still the best form of attack. It is vital to have a strong online presence, however, print materials bring in a whole other segment of customers and if the goal is to get as many customers as possible, balance out your tool set.

Chapter 7

⸎⸎⸎

Analog Materials

Ladies and gentlemen in the tech space . . . I have an important announcement to make: Printed marketing materials are not extinct. They have not gone the way of the dinosaur, the black and white television set, and the iPhone 4. Now, more than ever, it's imperative to be well versed at digital marketing. However, don't forget the print side. For many people, a physical leave-behind is still an incredibly effective tool.

I meet hundreds of people in the industry every year and, taking into account that I've been helping tech companies scale and grow for 20 years, it's fair to say that I've met thousands of people in the tech space. Some of the younger people look at print media as old school thinking. That doesn't surprise me. What surprises me is when they can't figure out how their amazing product isn't selling with all the great SEO, social media marketing, and targeted ads they've paid for. There may be a time when printed assets such as brochures or business cards aren't effective any more, but based on the current landscape, that day has not yet arrived. In order to do business today, and for the forseeable future, you need some printed materials. There's no better place to start than with a brochure.

Brochures

Every company needs a brochure. Whether people come to your store, or you have outside sales people or manufacturer reps, or if you go to trade shows; a brochure will continue to speak for you long after you meet a prospect. During personal interactions such as a meeting or talking to someone on a trade show floor, there is too much going on for most people to make a buying decision. Some people really like the physical marketing pieces and it's in your best interest to talk to those potential customers in the way they prefer. Leaving people with a professionally created brochure gives everything you said about the company/products credibility. Here are some tips to make that brochure effective.

First and foremost, it needs to be visually compelling. Design matters. The color scheme should mirror your logo, website, business cards, and all other marketing assets. The front cover is not the place to plaster the CEO's face unless he or she is a bona fide guru. The cover should compel the reader to open the brochure. And, you only have about three seconds to make that happen.

Now to the important stuff, the copy that actually sells your company or product. This is your opportunity to tell your story and differentiate yourself. Tell – don't sell. The most boring pieces we've all seen are full of marketing jargon. We all know the words:

dynamic, cutting edge, amazing, and my all-time favorite, if you act now, you'll get two. That may have worked in the 1960's yet now today's consumers are impervious to this type of sales push. They don't want sizzle, they want steak. It doesn't convince anyone to do anything. You need to develop that unique selling proposition to connect with your target market. In short, your brochure needs what we call your "social pitch."

Social Pitch

In the world of sales and marketing, success doesn't happen through blind luck. You get what you pitch for and you're always pitching (a mantra I learned from Topher Morrison). In the world of technological products, some companies think that just features and benefits are going to set them apart from the plethora of competition. It's not.

There are many pitches you should have memorized and none is more important than a social pitch. A social pitch has four elemental components:

1. Clarity

2. Authority

3. Problem

4. Solution

1) Be succinctly clear on what your product or service is. Forego using extra superlatives and just write down what it is. If

it's a projector screen, say it's a projector screen. It doesn't behoove a screen company to say they are a visual enhancement company. Be clear. Clearly communicate what industry you're in. Clearly identify who your target market is.

2) State your authority. Have your products tested off the charts? Have they won any awards? Is it version 2.0 of a great product? You need to establish your authority. Perhaps it's through how many customers you have or how long you've been in business or, if you're a younger company, through communicating why you are the expert at what you do.

3) What problem does your company and product solve? Your customers purchase your products because it solves a problem for them. If an HDMI cable manufacturer has the only product available to meet the new HDMI 2.1 spec, say that. Include that it performs refresh rates including 8K60 and 4K120, Dynamic HDR and increased bandwidth to 48G. Then, depending upon the size of your brochure, go into more appropriate details that discuss how that problem affects your targeted audience.

Identify the problem your product solves and illustrate the problem in practical ways that your customers can identify with.

4) Now we come to the pay off, the solution. How does your product help? As I mentioned earlier, don't spew technical jargon. Give an overhead view of how it works and the benefits your product delivers to your customers.

I know this sounds like a lot but in reality, all of this can succinctly be wrapped into a few paragraphs. Remember, you get what you pitch for, and you're always pitching.

Other items on a brochure

A brochure could also convey your process of doing business. An interested party may very well be more inclined to call your company or send an email if they know what to expect.

If it's a small company, perhaps a blurb about the CEO. If it's a large company and the brochure is about a particular product, perhaps a blurb about the company. It's all about the ethos, who are you and why do you do what you do? What's your why? Show your passion. This is another storytelling opportunity.

Ideally, you'd like to attract clients just like your favorite client right? To that end, include testimonials from your target clients. Share what it is they do and how your product has helped them. It will speak to like-minded customers that readily identify themselves with your dream clients.

Offer levels. If you have three different versions of the same product, offer the coach, business class, and first class. If you're a speaker manufacturer perhaps you offer Silver, Gold, and Platinum. Clearly articulate what the difference of each is. I've seen "professionally done" brochures that have listed three versions of the same product and they each have a different price, but never told me why.

Avoid a tri-fold at all costs. As you can tell, I'm strongly opinionated on this. Research has shown that the human brain does not find the vertical shape appealing. Opt for a square or rectangular shape for best results. Have you ever received an invitation to a festive occasion such as a wedding or baby shower? The cards are rectangular or square, aren't they? I like brochures to be the same shape because the mind correlates it to being festive, not foreign, like a tri-fold.

As I close on brochures, one last thing . . . just get it done. Always remember that having something is better than having nothing. Done beats unfinished-because-it's-not-perfect every time. Create a brochure. Have fun with it. Brochures are still a key piece into having effective sales, get yours done and be sure to put a digital version on your website.

Letterhead

When we a new client hires us, one of the first things we do is develop a corporate identity for them. It consists of the following: business cards, logo, email signature, and a digital corporate letterhead. Yes, letterhead.

While having your company letterhead isn't as vital as it once was, there are still occasional instances when one is needed and when the time comes, you'll feel more confident sending out content on a page that has your company letterhead.

I'm not going to get into the details of how to construct a letterhead, other than to say that it has to have the same look and feel as your other printed marketing materials. I will say this—don't spend the extra money to get letterhead printed and keep boxes of it in a closet somewhere. The days of feeding a letterhead into a Xerox machine are long over. Create a digital copy and distribute it to whoever needs it. When the time comes to print it out, I would recommend using a better than average stock of paper.

Digital and Print Advertisements (Ads)

I am going to get into details on strategizing for advertising purchases in an upcoming chapter, but for now I want to discuss the advertisements themselves.

Digital ads tend to be straight to the point. Mainly because there are size restraints and there isn't much room for copy, especially if on social media. The lack of space is fine though; you want to keep your message simple anyway. Simple, yet succinct. Make sure you link your ads to a landing page. A digital ad without a link is meaningless. If you don't have a landing page up yet, don't spend the money on a digital ad. It will just create confusion and frustration to those interested in your offerings.

Print Ads, although not as popular as in the past, are surprisingly still relevant in the tech space. MarketingSherpa's research from October 2016 ranked advertising methods based on their level of trust by consumers. Print ads were the number one type of ads the US Internet users trusted when making a purchasing decision – by a whopping 82%. That was followed closely by TV ads at 80% and mailed ads/catalogs at 76%. This made me rethink some clients' marketing strategies.

When it comes to print ads, clearly state your single message. You have about three seconds to capture a reader's attention. Be conscientious of how many products you put on an ad page. Thirty products on a page may work if you're selling used cars but other than that, less is more.

Remember, in all your ads, digital or print, they are representing, and a reflection of, your company. Take special care to make your message clear and concise. It is very important to stick to just one message per ad. Be sure you are consistent with your brand look and feel. Last but certainly not least, if at all possible, get a creative designer to develop your ads. Again, you have about three seconds to catch a reader's attention so make the visual appealing.

Selling on Amazon

Amazon sales aren't for everyone although it may have a place in your marketing mix. Amazon is a great place to blow out old inventory or broaden your customer base to boost sales. Companies that benefit most from selling on Amazon are those that are selling products that are unique to them, niche products and refurbished or used products.

If you are considering this sales channel, keep in mind that the account setup time can be months. You have several options on how you sell there. Using the "fulfilled by Amazon" method requires you to send inventory to their warehouse for shipment and Amazon will automatically make your products eligible for Amazon Prime free two-day shipping. Selling as a third party is another option with a whole other set of rules and you must earn the two-day shipping option with reliable service. Keep in mind that there is a plethora of rules and fees involved with either method.

If you do decide this sales channel makes sense, it's imperative to have good customer reviews to support and drive sales. Video and media reviews are strong tools for this outlet. Conversely, your Amazon reviews are also a strong tool to incorporate on your website product pages via various review aggregations tools available. According to Search Engine Land, 88% of consumers trust online reviews as much as personal recommendations. That's a statistic that can't be ignored.

To App or Not to App – That is the Question

With the propensity of mobile devices in use, an app for your product and services is another marketing tool to consider. An app for your business helps you reach many more customers than a traditional website. If your target customer is a younger audience, chances are they are performing their searches on a mobile device and an app will help that prospective customer find you during generic searches.

App development can be costly though. If this is a tool you need and you're working with a limited budget, consider starting with a basic app to better manage the initial expense and then build upon it as budget allows. If you plan the app in advance of the development process and have most of the app content developed, that will greatly reduce the development price. Cost will be dependent on platforms, functionality, sales, and design.

If you aren't interested in an app or budget doesn't allow for it, be sure that your website is very mobile-friendly to create a good user experience. Google will also penalize you on searches if your site is not mobile friendly. When developing your website, be sure to make it mobile-friendly. You can see if Google considers your web page mobile-friendly by using this test: http://bit.ly/mobile-site-test.

Marketing Automation, CRMs and Artificial Intelligence

Marketing automation tools have been around since the early 1990's. Initially used for email campaigns, the industry has grown from $225M to over $1.65B in just the past five years. Marketing automation is used to increase qualified leads and sales productivity while reducing marketing expenses—primarily a marketing focused tool. Paired with a CRM, a sales focused tool which is used to help track sales activities from lead to opportunity to customer, they are a powerful tool for any company.

There are many CRM and automation tools available that range from free, like the basic version of Mail Chimp and HubSpot, to the very sophisticated SalesForce. I believe every company should use a CRM/marketing automation tool to help manage their marketing and sales processes. To select a program that will work best for you, consider the cost, what features you really need and the ease/difficulty to implement. At Marketing Matters, most of our customers use Hatchbuck, HubSpot or Zoho. These programs provide the ability to start simply and grow in sophistication as their sales and marketing needs grow.

Artificial intelligence (AI) is rapidly evolving and will quickly change the marketing automation landscape in the not so distant future. AI leverages historical data and applies what is learned to current contexts to make predictions. Most of us are already using it on a nearly daily basis with Amazon Alexa, Google Now, Apple Siri and Microsoft's Cortana. Amazon currently uses AI to provide recommendations to you based on your preferences, history, and interests.

AI is going to create more intelligent marketing campaigns with the ability to append missing information on data records and clean lead information. Once the clean lead list is available, algorithms can create predictive lead scoring and intelligent lead routing to the best sales rep based on their expertise, work load or other parameters. Your drip or nurturing campaigns will become incredibly accurate with the highly-personalized ways to communicate with prospects based on data collected and the trends identified. I love data and the insights it provides. AI will be our next marketing game-changer.

Virtual Reality (VR)

Virtual reality is still in its infancy. It has really only started to come into its own in 2014 with Facebook acquiring Oculus, Google launching Cardboard VR and Samsung announcing GearVR. In 2016, Facebook launched Oculus Rift and Sony jumped into the fray with Morpheus for PlayStation VR. If you've used any of these products, you understand what an impactful marketing outlet it can be.

The immersive experiences you can create between consumers and your brand are powerful. VR is already being used in tourism, cruise ship marketing and hotel marketing. President Obama used it to provide a virtual tour of the White House before leaving office.

We're just at beginning to see how VR can be used in tech marketing. The New York Times has been a true pioneer in VR storytelling and sharing the content they develop with their loyal readers – along with a complementary pair of Google Cardboard viewers. Companies are promoting their content for home use, on cell phones, in-store experiences and at special events. Some useful applications I've seen include:

- Demos that include full functionality of the product

- Bringing your storytelling to your audiences with an extra dimension

- Branded entertainment experiences

- Creating an aspirational lifestyle with your brand

VR content can be developed using a 360-degree video camera, with 3D animation or a combination. The video is the most economical way to approach this as all you need is a good 360-degree spherical camera. Imagine the possibilities and the impact you could create.

PART 4:

AMPLIFY

Chapter 8

❧❧❧❧❧

Let Them Know You

People do business with people. To take it a step further, people do business with people they feel they know, like, and can trust. Ask any networker and they'll swear by those statements. A big issue for tech companies is how to make consumers feel like they know, like, and can trust their brand? After all, this isn't one-on-one networking we're talking about. Ideally, you want to totally dominate your market. Understanding how to leverage public relations goes a long way for a company's success.

Public Relations

Public Relations (PR) is one of the hardest to understand concepts of marketing. Many executives consider PR to be "free advertising." However, it is neither advertising, nor is it free. Most marketing professionals understand that PR is a time consuming and labor intensive process. Yet, done correctly, it can give your business the best return for their dollar. Instead of telling you what to do, I'm going to pull the curtain back a bit and let you see how we handle PR.

At Marketing Matters, we don't just think up great strategic public relations plans. We execute them with discipline, precision, and fervent passion to see our clients succeed in fulfilling their mission.

When our client's company gets mentioned in the media— whether as a company profile, a positive review, or an interview with one of their representatives—they not only pick up the exposure, they also pick up tacit endorsements. A third party that the audience trusts is presenting you as a skilled, honest, and reliable expert. When you put out an ad, regardless of your intentions, it comes across as self-serving because it is. Yet, when a credible third party endorses you, it comes across genuine and authentic.

That's golden.

Those placements don't happen by magic, though. A successful PR effort is a full-fledged marketing campaign in itself—to the media. Our public relations team works to position our clients as the "go-to" source for expert commentary, or to place their product for review or favorable mention, or for inclusion in holiday gift guides, buyers' guides, and the like.

Everything we do we do strategically, whether it's a single press release, or an entire PR campaign. We also assist with strategic analysis, direction setting, and developing implementable action plans that optimize resources, engage all stakeholders and employees and, above all, are relevant to the organization's vision and mission.

We work with our clients to perfect their message and then match that message with selected, high-payoff media outlets by getting the right angle on the right message to the right media at the right time.

Yet media placement is just the beginning of public relations services. Additional services typically offered under a public relations banner also include:

- Speaking engagements

- Interview opportunities

- Press conferences and events

- Media and spokesperson training

- Product launches and reviews

- Media monitoring

- Investor and employee relations

- Case studies and other content development

Most companies can benefit greatly from public relations activities to build awareness in the consumer and B2B channels. Press coverage received is frequently mentioned by customers and increased website traffic or direct inquiries help verify the effectiveness of your campaign. Your word will be out. And, in the event you weren't certain, that's the whole purpose of marketing.

Award Submissions

Winning awards give you instant credibility. There's nothing like having your industry peers provide the endorsement that you have the best product, service or project out there. Awards speak volumes to your customers and prospects.

We've created hundreds of client award entries over the past 20 years. I've also served as a judge for CTA's TechHomeMark of Excellence awards that are presented each year at CES. Based on those experiences, I want to share some tips for creating winning award entries.

- Plan ahead and make the time to enter. Like the lottery, you can't win if you don't enter. When judging, I've seen product categories with only one entry. Guess who wins?

- Only submit one product or service per category. For example, if you submit two products for Audio Product of the Year, you are cannibalizing your own entry. Don't compete against yourself. Enter the products in different categories or if possible, enter them together to make a stronger entry.

- If you are a manufacturer or distributor, work with your customers to submit project awards. Integrators many times aren't aware or don't take the initiative to submit their outstanding work. If you know about a newsworthy job your customer is working on, be sure to help get photos and other information for the award and send in an entry for both you and your customer. Everyone wins with this approach. It's a great way to give your customers the recognition they deserve for creating these stunning technology marvels.

- Did I mention planning? Know what awards make sense for your company and plan those entries throughout the year. Make sure you have photos at a minimum. Some entries accept videos also. Know that ahead of time so you can prepare an award that wins.

- One of our company mantras is, "Why guess when you can ask?" If you aren't clear about how to answer a question or what category would be best to enter your product, service or project, ask.

- Answer each question uniquely. If the entry asks for benefits to the integrator and a product summary, don't cut and paste the same information for both responses. You typically will have a limited number of words per entry question. Use these words wisely.

- Make it easy for the judge to see why your product should win. It's fine to use bullets instead of a paragraph of complicated specifications to organize your information. It's much easier to absorb the content this way.

- Use the early bird entry dates when possible. This will save you money that you can use for the award celebration when you win.

When your company wins an award, I recommend several ways to promote it. First, make sure you spread the word internally. Everyone at your company had some hand in creating this opportunity for the company. Be sure to celebrate it with your employees. They will appreciate the recognition and knowing that they are winners. After all, everyone wants to work on a winning team.

- Next, be sure to add it to your website. Not buried on the third level of a product page – add it prominently on your home page, preferably 'above the fold', the top part of the page you see on your screen when landing on the site. If you have sliders/banners on your site, this is a great way to make it front and center.

- Update your company email signature to include that award. Yes, even the folks in accounting should help spread the good news. The award credibility is also meaningful to your vendors and suppliers too.

- Next, create a press release and/or a blog post talking about the award. Share it via a customer newsletter, RSS feed, with appropriate editors and industry colleagues. It's a big deal and your customers, business partners and associates will be glad to hear the news.

- The award should also be shared on all of your company social media outlets. When we win awards, I also share it on my personal accounts. Depending upon how you position your posts, you can get several outreaches from the award announcement itself, receiving the physical award, celebrating with your staff, and so on.

- In addition, the entity who gives the award will also do promotion on its winners. This typically results in GREAT media coverage for your company.

As a tech company, do you know which award entries make the most sense for your company? Award entries can be very time consuming to write and expensive to submit. Some are well over $1,000 per entry although most are typically around $500.

Depending upon your product or services, here are some of the top tech award opportunities worth entering:

CES Innovation Awards Program recognizes two levels of honorees in 28 product categories, an Honoree which recognizes a product or technology that scores above the threshold for a specific category, or Best of Innovation which is bestowed to the highest rated product or technology in each category. There are sometimes multiple awards given in categories.

The cost for an entry ranges from $475 for a CTA member who is exhibiting at CES to $1,300 for non-CTA members who will not be at CES. Entries are typically due in the August/September timeframe for the January Consumer Electronics Show. Don't bother entering for this award unless you have a revolutionary new product or technology. Evolutionary products won't make the cut. The awards are announced in early November at the CES Unveiled press event in New York and heavily promoted before, during and after CES.

CTA TechHome Mark of Excellence Awards are for custom integrators/installers and manufacturers with installed technology products. CTA partners with CEPro and Electronic House magazines to manage and judge the award process. These awards are given during CES in Las Vegas each January and give great industry-wide recognition to the winners.

Submission are typically from August to November with the finalists announced in December. Project submissions for custom installer range from $100 to $150 per entry. Product submissions for manufacturers range in cost from $250 to $399 per entry.

Digital Trends Top Tech of CES Awards recognize outstanding new products at CES over a variety of categories from automotive to headphones and wearable technology. They will also announce a Best of Show award. To be eligible for this award, products must be aimed at consumers and make their formal debut at CES.

Entries open in October and awards are announced during CES. Digital Trends also offers several other trade show focused awards throughout the year that include Mobile World Congress, E3, and IFA. Other awards include Car Awards, Home Awards, and Best Products of the Year.

Electronic House Magazine's Products of the Year Awards recognize the best technologies, products and services for today's Electronic House. These awards are for manufacturers and service providers that cater to the connected home. Electronic House Magazine also offers Home of the Year awards for integrators. These are typically submitted around the start of the year.

InfoComm Awards highlight the important contributions made by industry individuals to the science of AV. These awards focus on the awe-inspiring people behind the gear who make the magic happen. Entries are due in March and winners are notified in April.

CEDIA Best New Product Award is a coveted prize for the residential integration community. Manufacturers need to enter by May and the awards are presented at the annual show in September.

CEDIA Home Technology Professional Awards and Industry Recognition Awards are bestowed upon the top of the home technology providers. These award recognize projects completed and contributions to CEDIA channel.

CEPro Best and Commercial Integrator Best Awards are manufacturer awards honoring new products at CEDIA (CEPro) and InfoComm (Commercial Integrator). Entries are typically due 30-45 days before the show and award at those respective trade events.

Social Media

Social signals (from social media outlets) and engagement are ranking factors that influence your site SEO (search engine optimization) directly and indirectly. When a website visitor is referred from a social media platform to your site, search engines monitor how long they are on your site, and if e-commerce is offered on your site, they monitor if the visit converts to a sale.

It's important to consider that social media and blog posts show up in search results making it imperative that the content is engaging and relative to your business. Search engines are placing more importance on social platforms, looking at shares, likes, and overall interaction from sites like LinkedIn, Google+, Facebook, and Twitter. Engagement has become a more important factor in recent search engine algorithms. You may not see the direct sales effect from social media yet it matters to the search engines and your customers. Interaction with customers across multiple social platforms is ideal. You want to tell your story where your customers are.

There are several ways to start building a following on each of your social outlets. On Facebook, start by inviting relevant people to follow your company page. You can also build your audiences by promoting your pages to targeted audiences. A small budget can go a long way in building your audience and the targeting capabilities of both Facebook and Twitter promotions are tremendous. Engaging content, especially when shared, also help get new followers.

Your social media messaging should follow your marketing messages – with a few fun things included as well. I recommend creating a monthly calendar to schedule posts each month. Be sure to tell your story and include content like your latest product information, tips and tricks for using your products, media coverage, events, blog posts, industry news (as relevant), and the like to the schedule. Continue to add material to the schedule throughout the month as you see fit – a breaking news story, images from events you attend, etc. Using scheduled posts ensure that the content can be reviewed and modified as needed.

Social media activities must be monitored continuously with timely and appropriate responses, including posts and private/direct messages sent to you. More and more customer service issues are flowing to social media and you will be judged on how and when you respond. Don't ignore a legitimate customer issue.

Be sure to keep your social media page clean. Hide inappropriate comments and sales pitches. Block the pests yet keep it real so your audience knows who you are and can respect how you handle an upset customer. None of us is ever perfect and your response will say much about your company and its values.

For most tech companies, we recommend using Facebook, Twitter, YouTube, and LinkedIn. I also highly recommend using Google+ for its SEO benefits. A Google+ post with lots of +1's will rank higher than most content you have out in the world-wide web. If you have consumer focused products, Instagram, Pinterest and maybe even Houzz could make sense for you.

Trade & Consumer Shows

Exhibiting at trade shows is the most economical way to get in front of several hundred or several thousand prospective customers. At most trade shows, the general public is not allowed entry. So, you won't be wasting time with potential consumers who aren't interested in your product as a car salesperson does with tire kickers who want to test drive a car they will never buy. The people who go to these shows, flying in from different parts of the country, and sometimes from different parts of the world, are the movers and shakers in your industry.

If you sell your products any way other than consumer direct, you will likely need to participate in one of the industry trade shows.

Depending upon your specific product or service, there are a variety of trade shows available to reach the B2B customer you desire. Here are some of the more notable shows we typically will recommend that our tech clients participate in:

Consumer Electronics Show *(CES)* is the Consumer Technology Association's annual signature event held each January in Las Vegas. CES represents the $287 billion U.S. consumer technology industry and offers great opportunities to show your products and services to buyers from around the world. Yet, with more than 177,000 people attending the show in 2017, it's challenging to be noticed. Don't commit to CES unless you have a strong plan for success and a decent budget to support it.

Integrated Systems Europe (ISE) is a collaborative effort between InfoComm and CEDIA. The annual event is held in Amsterdam in early February. With more than 75,000 attendees and about 1,100 exhibitors, it's the largest systems integration show in the world. If your integration products are ready to be sold outside of the United States, this is the show to attend. ISE helps their exhibitors find international distributors, has a strong educational curriculum, and is one of my favorite shows to attend each year. By far, the RAI Amsterdam has the best convention food I've ever had.

The International Security Conference West is one of the biggest security shows of the year. It's held in March or April in Las Vegas, and features more than 1,000 exhibitors and brands. About 30,000 security professionals attend each year. This is the show if you're in the security business or if you have integrated products that can be sold and installed by security professionals.

CE Week is typically held in mid-to-late June in New York, and it's New York City's largest technology show. CE Week creates a merger of Fashion Week with technology at this annual event. It's a smaller show with only about 3,000 attendees, yet they do feature some of the hottest trends in tech with a focus on back-to-school and holiday season product launches. Attendees are a combination of press, buyers, analysts and tech influencers, including bloggers, VCs and financial institutions, schools, start-ups, tech-forward citizen journalists, seniors, and moms. Exhibitors can develop one-to-one relationships with their best potential customers and advocates.

InfoComm is also held in June of each year. InfoComm serves the pro AV industry and rotates between Orlando and Las Vegas as their location. InfoComm has almost 1,000 exhibitors displaying thousands of products. They reach 40,000-plus attendees from 110 countries at the annual conference and offers a strong education curriculum for the attendees.

CEDIA is the Custom Electronic Design and Installation Association's show focused on residential technology. It's typically held in September and the show location varies from year to year. Attendance is about 20,000 at the annual event with a strong focus on education.

CTA Innovate! and Celebrate conference is another event I typically like to attend each year. Held in the fall, it's not a trade show per se. It's an opportunity to network and connect with CTA members, tech startups, industry thought leaders, venture capitalists, investors and tech media. The programming is innovation focused and the keynote speakers are always fabulous. CTA builds in plenty of networking time between sessions and it's a powerful event that provides the opportunity to meet influential people and gain insight into the CE industry.

If you decide to have your company participate in an industry trade show, be sure you have specific goals and outcomes for the show established during your initial planning. Those goals will drive your exhibit activities. If your goal is to get 600 integrator or retailer leads, make sure you have sufficient staff and demo areas in your exhibit to have that goal come to fruition. Promote your new product information and show activities to the press to get on your prospects radar before the show. Send email invites to your followers and consider renting a list from one of the trade publications to promote your show activities. Hold a beer reception in your booth to start new conversations. There are many activities you can create to reach your goal.

And here's a good tool to help you with your planning and budgeting – our trade show budget worksheet: http://bit.ly/show-budget

Chapter 9

Find Your Prospects

Website SEO

SEO (search engine optimization) will ensure that your website can be found by potential customers. SEO "optimizes" the site so it will appear closer to the top positions in the search results of Google, Yahoo, Bing, or other search engines. Research shows that people typically look at the top five results. The order of how sites show up in a search is determined by complex algorithms, and finding the correct patterns that machines will read optimizes which sites will rank higher than others. Search engines, such as Google, etc., constantly update their ranking algorithms to give users the best results.

SEO is not Internet marketing in itself; it is but one tool so that search engines will give you higher rankings. It is important because a good SEO approach can drive more traffic to your website, blog, or online store and help you gain more customers, make sales, and fulfill your business purpose.

As the search engines continually update their ranking algorithms, there are several things that can be done to improve your search ranking.

On-Page Optimization

On-Page Optimization allows search engines to crawl your site and tells them, as well as your website visitors, what the pages are about. There are guidelines and techniques to make sure the title tags, meta descriptions, alt tags (picture tags) and the H1/H2 header tags are properly used to ensure visibility. The load time of each page is also an important factor with ranking. Loading time is typically associated with the hosting service and the size of the graphics used on the landing page.

Directory Listings

Directory listings, or citations, are an important part of a local search strategy. These types of links have proven to increase local search engine rankings dramatically as they provide credibility for your business. Search engines reward sites that go the extra mile to be authenticated. If your business is listed in directories for your local chamber, local business listings, or a city business index, these listings verify your business before they allow you to join the directory, thus giving that credibility. If a media outlet writes about the company, additional credibility is given.

There exists a plethora of citation listings, each with their own verification process. Select directory listings that are right for your business and feed consistent information to them to help the rankings.

Reviews

Reviews are another crucial element that goes into local search strategy. A good mix of legitimate client reviews on Google and sites like Google+, Mantra, and Yelp, and press product reviews in consumer or trade publications will go far to help search ranking results.

Content

Social media posts, essays, and blogs are becoming a central part of search results. With click-through rates from content influencing rankings, it's imperative to produce content that gives you credibility within your industry and is appropriate for your target audience. Having a content writing strategy in place is crucial to search results. When the content writing strategy is part of your overall digital marketing campaign, including the social approach, it allows for a cohesive strategy of all the efforts working together.

Inbound Links

Just a few years ago, link building was the number one way to rank above your competitors. That's no longer the case. Directory listings are a more effective way to bring quality links to your site. Other links that help with rankings include press releases, articles, and posts from magazines or newspapers and guest posts.

Powerful Presentations

At some point, you or another team member will need to give a presentation to a potential investor, customer or other audience. 30,000,000 PowerPoint presentations are made around the world daily. How many bad presentation have you been through?

Sitting through a long, boring, PowerPoint presentation that no one wants to listen to can be torture. It's bad enough to sit through a monotonous, un-informative presentation, yet it is a much worse scenario if you are the one giving that type of presentation. It can damage your credibility and ensure that you're not invited to present ever again.

We all like to be entertained. We all like to learn, well, at least I hope we all still do. To make a presentation memorable and powerful you need to present information that's relevant to your audience in an entertaining manner. That's a lot easier said than done, as is proven every single day in the hundreds of thousands of networking meetings across the United States.

Here are 16 tips to deliver a powerful presentation:

1) Who's Your Audience

Know who you are presenting to. Figure out, if you can, their average age, job titles, and what their responsibilities are. What are they expecting to hear from you?

2) Your Message Is Important

Try your best to tailor the message to your audience. I'm not saying to speak with a southern accent if you're from New York and you're speaking to a group in Savannah, Georgia. Know what topics will relate best to your audience.

3) Keep Your Slides Short and To The Point

Each slide should only be about one key point or take away. If you put too much information on one slide it will become confusing to the viewer. Try to focus on putting one point and then some bullets explaining it on each slide. Too much information will have them reading instead of listening.

4) You Don't HAVE to Use PowerPoint

If you don't have PowerPoint you don't have to worry! There are some great options that you can make a great presentation with as well, such as Prezi, Powtoon, Keynote or SlideRocket. These tools make it easy to create engaging and professional presentations.

7) Use Bullet Points Rather Than Paragraphs

When you use bullet points on your slides, it gives your audience key points from your presentation. It also gives you speaking cues, in case you lose your train of thought.

Stay in control of the room and do the talking, don't let your slides speak for you. Limit your words to 4-6 on each bullet point.

8) Don't Read Your Slides Word for Word

This is a pet-peeve of mine. Someone is presenting and he or she turns to the slide and reads it for everyone. Use the words on the slides to prompt what you're going to say, not to say what you're going to say. Be sure to practice your presentation so you aren't held slave to the slides.

9) Use Easy to Read Text

Use text that is easy to read for your audience when you are giving your presentation. Don't use fancy fonts. Go with simple fonts such as Ariel or Times New Roman. Also, in regards to the font size, I would not recommend anything smaller than 16. Give a chance to the people in the back of the room to read your slides.

10) Bring Your Own Hardware

Don't depend on your host to have everything you're going to need. Make sure to bring your own laptop and your presentation in a USB flash drive. For an added measure, you can save your presentation in DropBox.

Also, make sure to bring the proper cables/cords. I've seen very confident speakers lose their cool because they didn't bring the proper hardware.

11) Try To Keep It Under 20 Minutes

Even if you have forty minutes, shoot for 20. There are studies stating that the average attention span of an adult in a presentation setting is 20 minutes. If you're going to go longer, do something to get them out of their seats and get the blood circulation flowing so they can regroup, reenergize and reengage.

12) Use Videos to Add Engagement

Sometimes a video will add a little something extra to your presentation. Use them when they are beneficial. Communicate your idea in a 1-2 minute video. Ensure that the production value of the video is good. The video should directly relate to your presentation. Ensure that you are only using videos in your presentation to add value.

13) Quality Over Quantity

Don't waste your audience's time with fluff and nonsense. Use your slides for quality information and only use them to enhance your presentation. Less is more. The more words and fluff on a slide, the more chances your audience will have to tune out. You don't want that.

14) Practice, Practice, Practice

Preparation is the key to giving an effective presentation. So, write a complete outline of your talk in bullet point detail. Next, dictate your talk into a voice recorder or cell phone, and then listen to it. It's amazing how much different it sounds when you hear your own voice. You'll hear ways that you can present the content differently.

Sometimes just changing the order of points increases their

impact.

15) Learn From The Best Presentations

Some presentations are better than others. Look at other presentations for ideas before creating your own. Keep an eye on the way they use of visuals and layouts.

16) Make It Actionable

Provide something at the end of your presentation that your audience can do immediately to take action. This is the exclamation point at the end of your talk. It's where you wrap everything up and bring everything together. What can your audience do when they walk out of the room to put what they learned into action?

Video Distribution

As mentioned earlier, videos are a strong tool to communicate your story, product information and a plethora of other topics. Once your video is complete, be sure to share it on Vimeo, YouTube, and your social media outlets. Add it to the product or support pages of your website and send it out in your customer newsletter. Like any other content, the key is to share.

And, don't forget about your live video tools available on many social media outlets like Facebook, Instagram, YouTube, Twitter, and Google. Live video is a good tool for personal interaction. It lets you be real and lets people get to know you. Live broadcast your product launches and trade shows. Address blog comments, give an inside look at your business, promote that upcoming event and answer

FAQs. There are many practical uses for live video. Once created, be sure to share as you would with any other video content.

Email Marketing

E-mail marketing is still a significant tool to use with your key audiences – B2B and/or consumers. You control the message and who sees which message ensuring delivery of relevant content. With the right frequency, it's also a strong tool to keep your company top of mind with customers and prospects.

Interested subscribers should be able to sign-up for the mailing list on your website, via social media links, and from campaign landing pages to help build the list. List purchases are another option. List data has become far more specific in recent years. Demographic and psychographic attributes are extremely targeted and this provides us a manageable way to test messages, then using those results, create larger campaigns.

It's important to manage your lists to maximize their use. We recommend setting up a CRM before going too far down this road. As mentioned in Toolify, options range from free simple list managers – for instance, Mail Chimp or the free version of HubSpot to customized SalesForce CRMs and everything between.

Elements of an e-mail campaign include:

1. Designing an email template for each audience

2. Creating the content and newsletter layout for the initial outreach

3. Sending the e-mail or e-newsletter (A/B testing of headlines is recommended)

4. Monitoring and reviewing campaign analytics to fine tune headlines, messaging, and design to increase open rates

I strongly recommend you engage in regular e-newsletter communications to both consumers (if applicable) and your B2B resellers. I'm a big fan of repurposing and like to create a blog post for each audience, post on your website, and use the content in a monthly e-newsletter. Other content for the newsletter can include event plans, product launch information, announcing new retails partners, testimonials, product reviews, and other relevant company information. The updated content on your site also helps improve search engine optimization rankings.

Be sure to include important links, a calendar of events, videos, calls to action and other marketing tools in your newsletter to maximize its effectiveness.

Direct Mail

With the inbox overload most of us deal with on a daily basis, you need to consider direct mail as an option in your marketing mix. It gives you an opportunity to provide a personal touch in your communications and talk directly to your prospects. And, as earlier mentioned, direct mail is one of the most trusted marking tools you can use.

Direct mail should be sent to a very targeted list to minimize waste. It's easily integrated into campaigns that include web, media and social to ensure measurement and accountability. It's tangible and allows your customer to interact using coupons, QR codes and personalized URLs.

There are so many tactics you can use to make sure your direct mail piece is opened. Color envelopes with a striking design and handwritten font in a lumpy mailer will virtually guarantee your package will be opened. And, 98% of folks open their mail on a daily basis.

Ad Buys

Even as a start-up with a limited marketing budget, some part of your budget will eventually go to advertising. A trade publication may call you with a pre-show new product launch program that too good to pass up or remnant space in a key tech publication is available at a blow-out price.

The ad placement exists to help you boost brand awareness, loyalty, promote events and product launches and ultimately, help you sell more stuff. Remember that regardless of the type of ad, without good creative it has little chance of success. Here are some useful types of ad purchasing that you should consider:

Digital Advertising

This is also called display advertising. Depending upon your specific marketing verticals, go to market strategy and budget you can use a variety of tools from product/service specific and abandoned shopping cart ads to remarketing and incentive ads. The topic of display ads area whole other book. Keep in mind that the average click through rate is only 0.17% - not a huge return on investment. Yet, here are a few display ad tactics that may make sense for your tech company.

Pay Per Click Campaign (PPC)

Be sure to set specific goals for the PPC campaign, which will likely include branding, lead generation, and conversions. Each of these goals will have specific cost goals measured in cost per action (CPA). Target the CPA to be lower than the product cost to prevent operating losses or at whatever level you believe is appropriate to gain market share. Make your ads creative and keep your ads fresh to prevent response declines.

AdWords

AdWords account should be set up using descriptive names for the campaigns and ad groups, making sure the campaign targets on one product category. Each ad group typically has four ad sizes and target a specific product or keyword theme. In addition, the campaign settings need to be optimized to align with the business goals—the locations, languages, network, and devices that are relevant to the target market.

Use conversion tracking to see what actions occur once a user clicks on the ad. It's important to know which keyword, ad, ad group, or campaign triggered the event so that you can adjust budgets to the higher performing campaigns. Also, be sure to address negative keywords to prevent unnecessary budget spending. Use AdWords customized alerts and automated rules to manage and monitor the performance of the campaigns. Google has some great tools readily available to help manage AdWords campaigns using their Keyword Planner.

As with any campaign, testing, measuring, and adjusting is critical to long-term campaign success. What works today may stop working thereafter. Ads, keywords, and landing pages should be optimized throughout the campaign as needed.

Ad Retargeting

Ad retargeting can be a powerful digital marketing tool. With retargeting, the brand can become more visible to those who have visited the website and help convert those visitors to customers.

Just because someone visits a website doesn't necessarily mean that prospect wants to see that company on every site they visit after. Using frequency caps, you can minimize overexposure that can lead to decreased campaign results. A frequency cap limits the number of times a tagged user sees ads to prevent them from becoming "banner blind" (ignoring ads completely) or developing a negative impression from overexposure. It's all about creating balance.

When e-commerce is available on your site, include another best practice—a "burn pixel." Burn pixel is code that is placed in the post transaction page to un-tag visitors who have made a purchase. This stops you from sending ads that simply waste budget. If you do want to retarget converted customers, you would do so with a different campaign.

Retargeted ads should serve the user based on their stage in the purchasing process. Using different retargeting codes based on the pages visited, the visitor receives an ad appropriate to their depth of engagement. For instance, a home page visitor would see general brand awareness ads when visiting Facebook or other sites. Someone who visits specific product pages would receive a specific product ad. This audience segmentation ensures you provide relevant and engaging ads at the right time.

Print Ads

For print ads, be sure you are selecting the right publication to reach your prospects. Get their media kit and examine the circulation and reach of the publication. Review the profile on readership and their editorial calendar. Is it a good fit?

Size and ad placement will depend on budget and content. Typically, next to the back cover, the right-hand pages in the first third of the book get the most attention. Yet, it may make sense to place your ad next to the editorial feature that supports your product category and messaging.

Although you can't accurately measure the impact of a print ad like you can with digital ads, you do have the advantage of more highly targeted audiences, a longer shelf live and a fairly captive audience.

Advertorials, Sponsored Content and Native Advertising

With over 2.7 million blog posts written and published daily, it's becoming increasingly difficult for marketers to get their content seen. Social media networks like Facebook and Twitter are constantly adjusting their algorithms to minimize organic content for brands. Banner advertising is less and less effective as people become banner blind. According to Solve Media, you're more likely to survive a plane crash then click on a banner ad.

Sponsored content or native advertising is paid advertising that contains the look and feel of the platform it's being displayed on. It looks like editorial content that should be there. It many times gets more attention than a display or print ad. There is some controversy over this in that it appears the publisher is creating a conflict between editorial and advertising. For instance, on Buzzfeed's website, the only content shown is based on native advertising. Readers don't always know what is editorial and what is advertising. Although this sounds like a marketing dream, it can create a conflict for readers who feel duped and does diminish the media's credibility. Yet, with advertising revenues declining for many publications, they need this tool to stay afloat.

If you do use native ads in your mix, make sure to use language that communicates that the content is paid advertising and that it is visible enough for readers to notice it on the page. Of all the advertising options available, this is likely the most effective.

PART 5:

TWEAKIFY

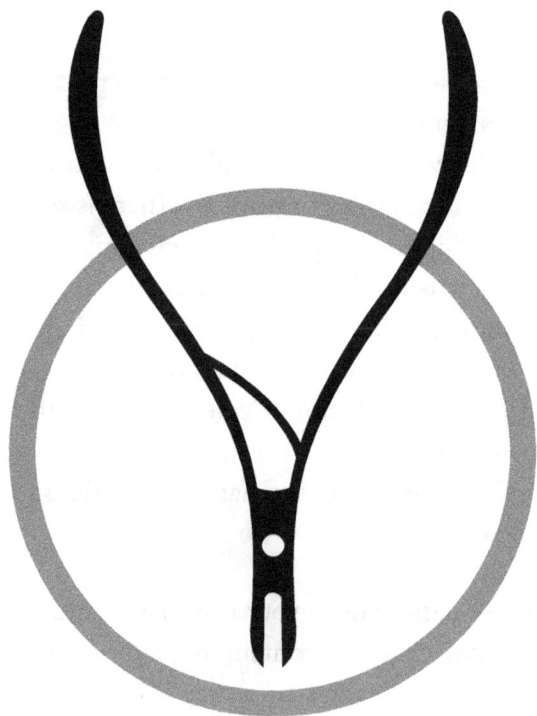

Chapter 10

‿◦❦◦‿

Test to Improve

Test to Improve

I'll never understand why tech companies don't test and retest their marketing endeavors like they do with their products. Throughout the process of developing a new product/technology, there are checks and balances at every turn. By the time the product is ready to be sold, it has gone through a series of tests. The ones it has failed has sent teams back to the drawing board until they got it right. Tech companies are familiar with this. The ones that aren't are either no longer around on their way out of business.

Yet, when it comes to spending their money on marketing, they take a much different approach. Executives get presented an idea (or ideas) they feel could work and they pull the trigger. I don't believe in a "let the chips fall where they may" strategy. It's suicidal for a company.

Every segment of a marketing program needs to be evaluated for its effectiveness. If a company uses print, social, ads, and television to market their product but don't have a way to determine which one(s) are actually bringing them customers, they're throwing good money away.

Oddly enough, many bigger or very successful companies, the ones with every resource at their disposal, don't test every segment of their marketing plan. They figure that because they're doing well, everything must be working. The reality of the matter could very well be that only 50% of their marketing plans are working out well and the other 50% are so bad they're doing worse than not bringing them customers; they're hurting their brand. If a company is doing well with 50% of their marketing efforts, they would totally dominate their market if over 80% of their marketing efforts were yielding great results.

I'll say this again because it's critical and could be the difference between success and failure for many businesses; every segment of a marketing program needs to be evaluated for its effectiveness. Yes, even when everything seems to be working perfectly. Do you recall the 100% rule? Nothing works 100% of the time and nothing works forever.

It might be time to refresh your thinking. Let's take your newsletter as an example, are you testing your headline? Do you even know how to do that? It's simple really, send half of your emails with one title and the other half with a different title. Now, determine which emails are being opened more. Voila, you've tested your headline!

Let's suppose you are offering multiple products and market

them together. Which offer is getting you more responses? If you have three products on a brochure and only one is selling, it might be time to replace those nonperforming products on the brochure and try some complimentary items that go well with the one that's selling.

Ads need to be tested and rotated as well to ensure you keep your users engaged. According to a recent study, click-through rates drop 50% after five months of running the same ads. Folks tune out what they have repeatedly seen. Refreshing ad materials is crucial to long-term success, however, if you never test it to find out when to change it, how will you know?

Always test and measure to make your money go as far as it can. When I talk to executives from start-ups, they'll tell me they can't afford to spend time and money on testing their marketing, to which I answer: "You can't afford not to."

Google Analytics

Google analytics is a freemium web analytics service offered by Google that tracks and reports website traffic. If you're wondering what in the world "freemium" means, it's a pricing strategy by which a product or service (typically a digital offering or application such as software, media, games, or web services) is provided FREE of charge but money (preMIUM) is charged for proprietary features, functionality, or virtual good. FREE - MIUM.

It's the most widely used web analytics service on the Internet. Naturally, you need a Google account to use it. It's a powerful tool. It can tell you which online campaigns bring the most traffic and conversions. It can also inform you of where your best visitors are located geographically. Imagine the strategy you could develop by just knowing where your best customers are located, which online campaign is bringing you the most traffic, and which online campaign converts the most people from tire kickers to users?

But that's not all. Would you like to know what people search for when they're on your site? How about the last thing they see before they leave it? How much better could you optimize your website if you knew what they are looking for and what made them decide to bail?

Understanding and leveraging Google Analytics, and other software programs like it, are vital to the growth of every company that wants to dominate online. Here are some of the Google Analytic features that warrant review.

Bounce Rate

We've all heard the term – bounce rate. Some people know it means when visitors leave your website and that's true, however, there's a little more to it. A bounce rate is also measured when a visitor stays inactive on a page for more than 30 minutes. I'm sure you've opened up a browser, thought about something else, and opened up another one, leaving that page open for hours, days, or even weeks. You plan on getting back to it but keep being pulled in different directions so it takes a long time to circle back to it. (Or is that just me?) Think about that when you get a report on the average amount of time a visitor is on your site. If they're not being active, they may not have left your site or closed the browser but they're not effectively on your site anymore, they've bounced.

When putting together a strategy that takes into account the bounce rate, this is the most important thing to know – bounce rates for first time visitors are far more important than the bounce rate of returning visitors. If people continue to visit your site, it doesn't matter much when they bounce.

The most important thing is to make sure you have a very clear call to action to get people to further engage on your site. Make it easy for your visitor to find it and give them an incentive.

To reduce bounce rate, you could have links to other products or blogs within your site. Studies show that the earlier the links are on a page the more clicks they get. Google loves the click-through. Google will be much friendlier to your website than your competitors if your average visitor is visiting three or four pages and the competitors' visitors are only visiting one.

Advanced Segments

Don't be afraid of the advanced segments on Google Analytics. They make it as user friendly as they can, although granted, some things, such as writing code, requires an expert, yet not all.

You find the Advanced Segments when you get to the Traffic Sources Overview section, it's directly under it in the first tab. There you can do searches for All Visits, New Visitors, Returning Visitors, Paid Search Traffic, Non-Paid Search Traffic, and more. You can also use the Custom section and enter in whatever you want, such as Social Media. The predefined advanced segments are great and I always like to accompany it with ones we customize.

Filters

If you have a large company, and every employee spends time on your website, your traffic will be skewed. The way to fix that is through the Filters section. Basically you add the company name to your filter, so if you don't want Google to add YOU as a visitor, you might name it - My Computer. You click on Predefined filter. Click on the Exclude feature and type in your IP address. You can do that for your entire company so that you have the most accurate data from real visitors.

Custom Alerts

You can set up alerts to warn you when things are going exceptionally well or exceptionally bad. As an example, if your average click through rate on emails is at around 7% but you want to be notified when an email gets to 15% conversion, set it up in the Custom Alerts section and you'll be notified when the email conversion gets to 15%. Or, if you average 300 visitors a month, you might want to know when you've gotten 500 or less than 100 visitors in a month. That way you can see what you were doing right and keep doing it or try to figure out why you had a massive decrease in visitors.

There are many features on Google Analytics that have proven to be extremely useful to our clients. I recommend you spend some time figuring out how to leverage the in-depth analytics you can get online. It can make a difference between being profitable or being wildly successful.

Surveys

There is probably no easier way to get feedback from your customers than to ask them to fill out a survey. The more popular ones are from companies such as Survey Monkey, Question Pro, and Zoomerang. Even Facebook has a decent version that allows you to poll. If you aren't conducting surveys, it's difficult to know which areas need improvement.

"Gathering feedback from customers has a great impact on any organization's business model. Feedback, when used correctly, can increase cross-sell and upsell transactions by 15 to 20 percent." Says, Esteban Kolsky, a research director at Gartner Inc. in Stanford Connecticut.

The main problem companies face when trying to get consumers or their top dealers to take a survey is that it's so one-sided that people don't take the time to fill it out. I've been in rooms where the marketing manager has wanted to pull his hair out because people aren't taking just ten minutes to fill out their survey. That's usually when I shake my head and realize that marketing isn't for everybody. Most people aren't going to take ten minutes out of their day without a return.

5 Simple Steps to Get People to Participate in Your Survey

1. Make it short. Two minutes out of a person's busy day is too

long. I've seen surveys with over 30 questions. You need to consider that people value their time and if you require too much of it, they will not give you any of it.

2. Get right to it. Have you gone to take a survey but had to fill out an in-depth questionnaire first? They want your age, sex, race, birthdate, email, phone number, city, state, and so much more. By the time people are done filling out the questionnaire, they won't take the time for the survey. You don't need that much information. Name and email address works fine.

3. Give people a reason to take it. If you ask people to take time out of their day to take a survey, what's in it for them? Offer them something, make it worth their while. A "Thank You" is nice, but a discount is better. The other day, I called a friend who was literally taking a survey from the place he had bought his car. We had an urgent matter to discuss but he asked if he could call me back. When he called back, about a minute later, he informed me that by taking the survey, his dealership offered him a free detail the next time he got his oil changed there. We've had great success with surveys at trade shows by requiring the short survey completion to get the fun promo good.

4. Make it Multiple Choice. Some people don't have the time to write out the answers to your questions, particularly those that take it from a mobile device.

5. Allow people to add additional comments. To take a survey from 1 to 5; 1 being terrible service and 5 being excellent

service, will only get you marginal information. For those that want to take the time to tell you why they are voting the way they are, will get you more information than the multiple choice setting.

The good thing about getting people to take surveys is that it's usually free. If you want to get more intricate you may find yourself paying $20, still, very little money for valuable information. Invite your top prospects, customers, or dealers and fine out how you can fine-tune things.

Net Promoter Score

Brand Loyalty. Brand Loyalty. Brand Loyalty. If I hear that cliché one more time… However, although the term continues to be overused and abused, customer loyalty is a powerful thing. That's when a Net Promoter Score (NPS) comes in handy. This is a little different than a run-of-the-mill survey.

NPS is a management tool designed to gauge the loyalty of a firm's customer relationships. Fred Reichheld first introduced it in his 2003 Harvard Business Review article, "One Number You Need to Grow." It measures the loyalty that exists between a provider and a consumer. The NPS is based on responses to a single question: How likely is it that you would recommend our company/product/service to a friend or colleague? It's much less detailed than a regular survey yet it yields you great information.

More than two thirds of the Fortune 1,000 companies have been reported to use the metric. However, I think every company should use it, not only for their customers but also with their employees.

Many marketers out there believe that information is key. I don't necessarily agree. I believe that knowing how to get the right information and then knowing what to do with it is key.

Chapter 11

As you have read, marketing effectively today is not an easy process. There aren't any quick fixes. If you're struggling to get a good product or your company name in front of your top prospects, I don't intend to be crass, however, I only know how to say it one way, your marketing efforts are to blame. There isn't just one aspect of marketing you can incorporate that is going to get you from obscurity to a household name. You need a well-balanced attack.

"Marketing is a contest for people's attention." Seth Godin

The demand for people's attention today is a fierce battle. Livelihoods, careers, and people's futures are on the line. Agencies that have stood the test of time are still swinging away and inexperienced teenagers not yet old enough to go to college are building websites and starting their own "marketing" businesses. We have been called upon more than I care to recount to "fix" a website a neighbors kid did. Your reputation is too important to not get your message out right.

It's important to understand that consumer attention is limited; it isn't infinite. Now more than ever your marketing agency or in-house marketing talent needs to be vetted. Make sure you have an agency that understands your space, your products, and your customer. Even if you have full-time staff doing your marketing, it behooves you to have a relationship with a professional marketing agency that lives and breathes in your space.

The answer to the question: "In the midst of so many changes, how can I keep my company relevant?" has been answered in this book...for the most part.

The reality is that there are many more marketing tips, strategies and nuances within concepts than I could have written here yet I didn't want to write an encyclopedia-sized book. I've given you a solid foundation to ramp up your marketing efforts. The question now is what are you going to do with this knowledge?

Every vertical has its unique personality. For instance, the healthcare industry, the education industry, and the broadcast industry all have their distinct challenges and markets. Out of all the industries there are, technology is most likely the strongest vertical there is. Every industry I know of is dependent on technology. There is not another vertical in existence that permeates into many facets of every other industry like technology. I don't mean to come across as grandiose yet the future of mankind is dependent on technology... all of technology.

That being said, there is no better industry to be successful in than technology. Paul Lamkin, a contributing writer to Forbes.com, stated that the Wearable Tech market will be worth $34 Billion by

2020. That's right around the corner. That's $34 Billion. Wearable Tech is only one small slice of the technology industry. It has grown fast and the competition for people's attention there, like in every other area of technology, is furious.

I have been fortunate to have worked with and seen some great people with great products surpass their goals. As a marketer, the biggest compliment my clients have given me is that they have retained me over many years. There have been many laughs and quite a few disagreements along the way but for the most part, my one-time clients become my long-time clients. There's only one reason for that, we have proven to be worth the relationship.

I often travel to places across the country and the world to work the top tradeshows in the industry. I meet hundreds of brilliant people a year who specialize in marketing products and establishing start-ups. It is imperative for me to keep my finger on the pulse of the tech industry as well as the most effective marketing methods out there. And as I mentioned earlier, I've been a lover to tech all my life and have been a professional marketer in this space for over twenty years. I have seen many a fad come and go, taking with it many once-promising companies. I know that much depends on how your product or company gets marketed and that is why I wrote this book.

Much of what we do, I wrote for you here. I know that if my marketing strategy is used in its proper context, it will help. At the very least I hope this book gave you a few solid takeaways that you can implement right away.

I know some of you who don't have strong marketing

backgrounds and have to wear multiple hats in your business might be overwhelmed with all the to-do's I wrote. Don't get overwhelmed; you don't have to do them all. Pick the ones that seem to be a natural fit with your current skill set. Like a great movie, you first have to lay down a solid foundation and then build on it as you go.

If after reading this book, you are left with questions or uncertainties regarding how to market your product or business, please feel at liberty to contact me. Although my clients and my travel schedule keeps me quite busy, I wrote this book to be of help to those who aren't my clients and, as you took the time to read my book, I'll take the time to answer your questions as best I can.

The best way to reach me is via my email address: coleen@marketingmatters.net

"Good marketing makes the company look smart. Great marketing makes the customer feel smart." Joe Chernov

Lastly, I am honored and humbled that you read this book. I hope it will serve you well.

All my best,

Coleen Sterns Leith

Acknowledgements

Many thanks to Kyle Glass, Marie Holloway and Morgan Roush from the Marketing Matters team for giving me the time, help and support to make this book and the launch happen when time is always a precious commodity.

I'm indebted to Topher Morrison and his business coaching. Topher's processes made this book come together.

And, a big thank you to Eli Gonzalez for his dedication, input, and editing skills. Your help has been priceless.

Lastly, I am grateful to our clients over the past 20 years - without your trust in our solutions, this book would never have been written.

About the Author

Coleen Leith is president and founder of Marketing Matters, an agency working exclusively with small and medium-sized consumer electronics and pro AV companies. For the past 20 years, Marketing Matters has been the agency of choice for companies like Bose and Vivitek to launch new technology. Her team has helped take a startup company to $200 million in sales in seven years and has won 61 design and publishing awards for the work they've done.

Working closely with technology clients, Coleen and her team have created over 200 successful product launches and public relations strategies that have garnered national and international exposure and a dynamic market presence for each of their clients.

Coleen is an active member of the Consumer Technology Association's Audio Division board, participates in the TechHome board activities and is a CTA Mentor. Coleen is the past-chair of

CEDIA's Professional Services and obtained her CEDIA Certified Instructor credentials. She's a former board president for Big Brothers Big Sisters of Broward County, works with Big Brothers Big Sisters' national marketing committee and has been a Big Sister for over three years.

Coleen resides in Saint Petersburg with her husband and three French Bulldogs.

Recommended Reading

The E-Myth Revisited: Why Most Small Businesses Don't Work and What to Do About It

Oct 14, 2004   by Michael E. Gerber

The Dip: A Little Book That Teaches You When to Quit (and When to Stick)

May 10, 2007   by Seth Godin

Ninja Innovation: The Ten Killer Strategies of the World's Most Successful Businesses

Jan 6, 2015   by Gary Shapiro

Collaboration Economy: Eliminate the Competition by Creating Partnership Opportunities

May 6, 2014   by John Spencer Ellis and Topher Morrison